T0289571

World Class Master Scheduling

Best Practices and Lean Six Sigma Continuous Improvement

BY

DONALD H. SHELDON
CFPIM, CIRM

APICS.
THE EDUCATIONAL SOCIETY
FOR RESOURCE MANAGEMENT

Copyright ©2006 by J. Ross Publishing, Inc.

ISBN 1-932159-40-1

Printed and bound in the U.S.A. Printed on acid-free paper
10 9 8 7 6 5 4 3 2 1

Library of Congress Cataloging-in-Publication Data

Sheldon, Donald H.
  World class master scheduling : best practices and lean Six Sigma
continuous improvement / by Donald Sheldon.
    p. cm.
  Includes index.
  ISBN 1-932159-40-1 (hardback : alk. paper)
  1. Production scheduling. 2. Six sigma (Quality control standard). 3.
Production management—Quality control. I. Title.
implementation. II. Title.
  TS157.5.S52 2005
  658.5′3—dc22   2005022322

This publication contains information obtained from authentic and highly regarded sources. Reprinted material is used with permission, and sources are indicated. Reasonable effort has been made to publish reliable data and information, but the author and the publisher cannot assume responsibility for the validity of all materials or for the consequences of their use.

All rights reserved. Neither this publication nor any part thereof may be reproduced, stored in a retrieval system or transmitted in any form or by any means, electronic, mechanical, photocopying, recording or otherwise, without the prior written permission of the publisher.

The copyright owner's consent does not extend to copying for general distribution for promotion, for creating new works, or for resale. Specific permission must be obtained from J. Ross Publishing for such purposes.

Direct all inquiries to J. Ross Publishing, Inc., 5765 North Andrews Way, Fort Lauderdale, Florida 33409.

Phone: (954) 727-9333
Fax: (561) 892-0700
Web: www.jrosspub.com

Have no anxiety about anything,
but in everything by prayer
and supplication with thanksgiving
let your requests be known to God.

Philippians, 4:6

This book is dedicated to God for his blessings,
the most important of which are my family.

TABLE OF CONTENTS

PREFACE

This book is devoted entirely to master scheduling and related processes. The master production schedule (MPS) is covered in detail, including how high-performance companies design it, how it links to the top management planning process, the handshake between the demand and supply sides of the business (referred to in this text as the "rules of engagement"), and lessons learned around the world on this exciting topic.

There are very few well-written books on the topic of master scheduling. John Proud's book, *Master Scheduling*, which has been in print many years, is probably the desk reference on most of the better master schedulers' credenzas, especially if they have been in the business for several years. This book attempts a different approach. The objective of this book is to complement and fill in the gaps from Proud's reference text and bring best-in-class thinking into view, including very current topics such as lean and Six Sigma, topics that are on the minds of many managers as they work for competitive advantage.

Other texts dedicate page after page to the "math" behind promise dates such as available to promise. Modern enterprise resource planning systems today calculate the math automatically. Understanding it is necessary, but the roles and duties of the master scheduler are probably more appropriate areas of detail study today if competitive advantage is the goal. Although the "math behind available to promise" is included in Chapter 5, the real focus of this book also goes beyond the spreadsheet mentality of master scheduling and helps the reader understand the roles and responsibilities of the master scheduler in high-performance manufacturing today.

Master scheduling is a passion of mine and many others I meet in high-performance manufacturing environments, and for good reason. The MPS is a critical pivot point within the manufacturing world. The MPS represents the policing effort in the manufacturing environment, establishing and keeping required disciplines. Regardless of the geographic location where you manufacture, it has become more and more apparent that a prerequisite to good process and results is a disciplined MPS process.

From page one, this book is focused on bringing the reader closer to the experiences of real-life high-performance master scheduling taken from years of field experience and working with some of the best. Performance gaps are discussed, and approaches from centralized to remote master scheduling are compared. The first five chapters are dedicated strictly to good master scheduling technique and process, inventory strategy, and MPS application. Linkage to the supply chain and value-add supplier interaction designed to maximize critical advantage are detailed. Understanding the application of required "rules of engagement" and order management are discussed.

Chapters 6 to 12 deal specifically with the roles from schedulers to top management. These step-by step expectations are documented to help the reader achieve high-performance results in his or her respective applications. The last few chapters are dedicated to tools, management systems, and metrics. Software dependence and interaction are discussed, with arguments for robust process as a prerequisite to high-performance software tools.

Robust process requires a continuous improvement culture and mentality. Lean and Six Sigma as well as Class A ERP performance are threaded into the discussion in more detail in these last chapters, allowing the reader to understand the human issues and how some successful organizations have dealt with them. Expectations and criteria for achieving high performance are documented. The sales and operations planning process is also detailed, with schedules and even spreadsheet expectations that most readers will find helpful.

It was a thrill to be asked to write this book. I hope that many of those who have not yet acquired an appreciation for the MPS will be inspired by the examples in this book of high performance and good process. To help with the knowledge transfer, three interviews are also included in this detailed process description. The first interviewee, Bob Shearer, has been a master scheduler for several years in a heavy equipment manufacturing plant that does mostly make-to-order and engineer-to-order products. He is one of the most proficient master schedulers I have met and has been an inspiration in both my respect for the process and in the writing of this book. The second interview is with Jeremy Sauer, master scheduler in a mostly assemble-to-order and make-to-order business supplying retailers. The third is with master scheduler Robert Turcea. Robert is also a best-in-class master scheduler in a high-performance process but in a repetitive, commodity environment that utilizes mostly make-to-stock inventory strategy. These three diversified examples of good process with disciplined use of both the demand and supply sides of the business will be helpful in heightening understanding of why high-performance businesses hold master scheduling in such high regard. While these businesses have little in common from a market or product sense, the commonality of process discipline will hopefully become obvious.

ACKNOWLEDGMENTS

It has been a great experience to travel the world and both learn from some of the best and share with those in need of the knowledge. Consulting is the greatest job on God's earth. I have met so many friends and great people in my travels. The following people helped inspire this project:

- Anita Sheldon — My wife and inspiration, who has always supported me in every way
- Bob Shearer, The Raymond Corporation — One of the most effective master schedulers I have met (and I know a lot of them!)
- Robert Turcea, Grafco PET Packaging Company, Baltimore — For full cooperation in providing documentation and examples of good process
- Jeremy Sauer, Mathews, Inc. — For eager cooperation and quality in getting the interview chapter completed
- Drew Gierman, J. Ross Publishing — For asking me to do this project

A few others have helped in several ways to get this project together, some unknowingly, by providing opportunities, inspiration, good ideas, and/or process: Bill Amelio, Lenova; John Bertolet, PMI; Westy Bowen, E-Z-Go Textron; Patrick Burke, Mathews, Inc.; Rick Calkins, Honeywell; Michelle Dudley, Brennan Industries; Kathy Dyer, NCR; Tony Fraley, Electrolux Home Products; Tim Frank, Grafco PET Packaging; Neil Gibson, NCR; Jay Holt, The Raymond Corporation; Howard Lance, Harris; Mike Loughrin, Transformance Advisors, Inc.; Steve McPherson, Mathews, Inc.; Wendy Musson, Electrolux Home Products; Paul Potter, Buker, Inc.; Ron Sheldon, Grafco PET Packaging Company; Kevin Stay, Brennan Industries; Bob Truesdell, Electrolux Home Products; Tom Vinci, Stevens Roofing; Frank Wagner, Electrolux Home Products; and Bob Wilkins, Veeco Instruments.

THE AUTHOR

Donald H. Sheldon is president of DHSheldon & Associates in upstate New York. Past positions include Vice President of Global Quality and Six Sigma at the NCR Corporation in Dayton, Ohio; Vice President of Buker, Inc., a recognized leader in management education and training in Chicago; and General Manager of Aftermarket Services Division at The Raymond Corporation, a world class manufacturer of material handling equipment. He has more than 30 years of manufacturing management experience on four continents helping companies achieve business excellence.

Mr. Sheldon has written numerous articles in journals, is co-author (with Michael Tincher) of the book *The Road to Class A Manufacturing Resource Planning (MRP II)* (Buker, 1995), and is the author of *Achieving Inventory Accuracy: Class A Performance in 120 Days* (J. Ross Publishing, 2004) and *Class A ERP Implementation: Integrating Lean and Six Sigma* (J. Ross Publishing, 2005). He has been a frequent speaker at colleges, international conventions, and seminars, including APICS. He holds a master of arts degree in business and government policies studies and an undergraduate degree in business and economics from the State University of New York, Empire State College. He is certified by APICS as CFPIM and as CIRM.

ABOUT APICS

APICS — The Educational Society for Resource Management is a not-for-profit international educational organization recognized as the global leader and premier provider of resource management education and information. APICS is respected throughout the world for its education and professional certification programs. With more than 60,000 individual and corporate members in 20,000 companies worldwide, APICS is dedicated to providing education to improve an organization's bottom line. No matter what your title or need, by tapping into the APICS community you will find the education necessary for success.

APICS is recognized globally as:

- The source of knowledge and expertise for manufacturing and service industries across the entire supply chain
- The leading provider of high-quality, cutting-edge educational programs that advance organizational success in a changing, competitive marketplace
- A successful developer of two internationally recognized certification programs, Certified in Production and Inventory Management (CPIM) and Certified in Integrated Resource Management (CIRM)
- A source of solutions, support, and networking for manufacturing and service professionals

For more information about APICS programs, services, or membership, visit www.apics.org or contact APICS Customer Support at (800) 444-2742 or (703) 354-8851.

Free value-added materials available from
the Download Resource Center at www.jrosspub.com

At J. Ross Publishing we are committed to providing today's professional with practical, hands-on tools that enhance the learning experience and give readers an opportunity to apply what they have learned. That is why we offer free ancillary materials available for download on this book and all participating Web Added Value™ publications. These online resources may include interactive versions of material that appears in the book or supplemental templates, worksheets, models, plans, case studies, proposals, spreadsheets and assessment tools, among other things. Whenever you see the WAV™ symbol in any of our publications, it means bonus materials accompany the book and are available from the Web Added Value Download Resource Center at www.jrosspub.com.

Downloads available for *World Class Master Scheduling: Best Practices and Lean Six Sigma Continuous Improvement* consist of master scheduling training slides and a checklist for assessing your scheduling process performance.

UNDERSTANDING THE MASTER SCHEDULE

Master scheduling is probably one of the most important and yet least appreciated processes to control costs in business today. Top management ignores it, middle managers take it for granted, and line management often unknowingly subverts the integrity of the schedule by causing unnecessary changes to it. In contrast, high-performance businesses treat the master scheduling process as the heartbeat of the supply chain and utilize best practices in this space to minimize cost and efficiency and to maximize customer service. Techniques in master scheduling cover lots of business process territory, including links to demand planning, financial planning, inventory strategy, customer service, and rules of engagement.

Topics covered in the chapters to come include what good process would look like, how top managers interact with this driver of requirements called master scheduling, and how to implement a high-performance master production scheduling process in your business. Master scheduling is a passion with some manufacturing professionals. The objective of this book is to capture some of that passion. It is a pleasure to put into writing the power of one process that can help improve numerous processes within both manufacturing and distribution companies. Observation of hundreds of businesses using both good and bad process has helped with the conclusions. The logical place to start this discussion is by defining the master schedule.

THE MASTER SCHEDULE

The master schedule in an enterprise resource planning (ERP) business system is the detailed driver of requirements in manufacturing companies. This accounting of known and unknown requirements is a company's determination of firm and forecasted signals empowered to drive procurement and manufacturing requirements. This schedule can dictate amount of inventory and timing of the procurement and can even influence lot size or order size. This is an extremely important set of requirements and makes the disciplines and rules governing the schedule very important to most businesses concerned about cost, efficiency, and customer service.

In high-performance businesses, top management engages in a planning process referred to as the sales and operations planning (S&OP) process. It should be noted that some companies refer to this as the SIOP (sales, inventory, and operations planning) process. In this book, it will be referred to as the S&OP process. The master scheduler plays a critical role in the development of the spreadsheets and performance metrics evaluated in S&OP. Probably even more important is the requirement in high-performance companies that the master schedule link directly with this top management planning process. In the S&OP planning process, the CEO, president, and vice presidents of both demand and supply engage in risk management within the planning horizon of 12 rolling months. No one understands the risks of creating the right product using only forecasts for the demand signals better than the master scheduler.

In a high-performance ERP organization, sometimes referred to as a Class A ERP organization, the two most influential positions are the plant manager and, you guessed it, the master scheduler. The master schedule not only defines the activities required to meet the top management S&OP but also dictates the requirements from customer demand. This already sounds like quite a job and it is. Figure 1.1 depicts the relationship of the master schedule to the rest of the ERP process flow. Top management planning decisions feed into the master production schedule (MPS) and drive emphasis and create the forecasted demand. The MPS in turn feeds back vital data concerning customer activity and coordinates the interaction of the two inputs. This feedback affects the next S&OP cycle.

The MPS is the detailed schedule that resides within the ERP business system that drives inventory strategy, supply chain accountability, inventory levels, customer service, and machine and capacity utilization. The main categories that are driven fall into two areas: known requirements and unknown requirements. This may seem too simple to even state, but it is the essence of understanding the master scheduling process.

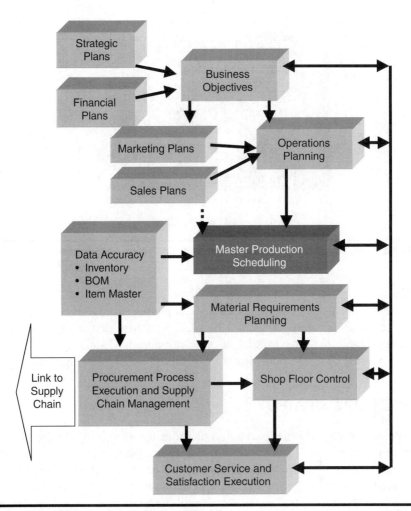

Figure 1.1. ERP Business System Model

The darker shaded areas in Figure 1.2 represent the known requirements. Where the darker shaded areas appear, there are normally many detailed stockkeeping unit (SKU) level requirements driven from either firm orders from customers or firm orders from planned requirements. In the case illustrated in Figure 1.2, if each rectangular shaded area represents one day, the conclusion from this figure is that backlogged orders only go out about six days, and even then capacity is not full for the six days. Only Monday and Tuesday are close to being filled (Tuesday might even be a little over capacity per the diagram).

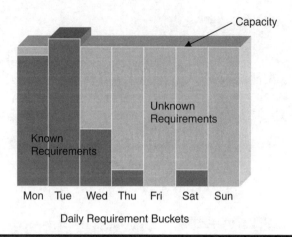

Figure 1.2. Master Schedule Known and Unknown Requirements

This is the basic structure of the MPS in its simplest form. As this book progresses, best practices examples of how high-performance companies deal with the realities of business will be used. The first step is to tie inventory strategy tightly to this schedule, in order to understand the impact. Inventory strategy must be well understood to execute master scheduling well.

INVENTORY STRATEGY

Inventory strategy (sometimes referred to as manufacturing strategy) is the buffer inventory plan for each product family in a planning process. The buffer inventory plan is the plan for where in the business the inventory will be buffered, if at all. Sometimes buffer will be in raw material, sometimes it may be in finished goods, and sometimes it will not exist at all except in the forecasted plan. If the flexibility and responsiveness within the supply chain match the need, there is no reason to have buffer. Very few manufacturing plants, however, meet the criteria completely to go buffer-free. In those companies without buffer, the plan has to be decided based on product family process capability. Figure 1.3 gives a sense of how these strategies fit together. Keep in mind that most companies do not engage in just one of the strategies. *It is most common for companies to have several inventory strategies, with separate pricing strategies, rules of engagement, and service requirements for each product family.*

It is necessary to review inventory strategy as you set up the MPS because the entire manufacturing planning world revolves around these strategies. Although many companies do not acknowledge this thinking, *all* companies have

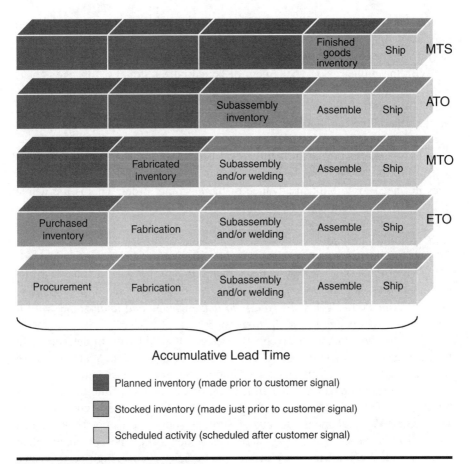

Figure 1.3.   **Inventory Strategy**

to make these decisions. Too many organizations will say they use one type of inventory strategy and actually do something totally different. Still others do not recognize that their strategies differ from one product family or line to another.

It is common for those in top management to believe that their business methodology is one inventory strategy. It is just as common for them to get involved with the process and find that they do not really understand what is going on in the detail planning process. Maybe it is unfair to be so critical, but most of the time, especially in so-called "make-to-order" environments, there are buffers distributed in various places within the process. These have to be recognized and understood in order to have real control over the manufacturing process. After all, these buffers are probably there for a reason. It is paramount to know what *is* in place and for what reason it was initially positioned. It could very well make a difference in planning methodology or inventory strategy overall. It certainly affects customer service and profitability.

With the proper acknowledgment of actual buffer strategy, management is more in control of the risks taken and no one is surprised by inventory that by design is *built* into the process. Inventory strategies change the method used for master scheduling suppliers and components. Inventory strategy is smart policy, but the rules agreed to internally — what products, how much, and when — need to be understood by all the players. Many times, finished goods inventory is scrutinized, as it is very visible and better understood. Sometimes inventory does not need to be built in advance of the customer order. This is the whole point of inventory strategy. Lean philosophy has gotten a lot of press recently and deservedly so, but inventory strategy has been around for years and covers some common ground.

It is important to realize that acknowledging rules of engagement and utilizing good inventory strategy is not a methodology designed to say no to the customer. *Inventory strategy is acknowledged so that everyone in the business has the same shared goals. Its role is to acknowledge and understand what it costs when we say yes to the customer.*

The acknowledgment of inventory strategy is not difficult to determine. In fact, it is dictated to you if the background data are understood. When lead time requirements from the customer are shorter than the total accumulative lead time of the supply chain and manufacturing process, some of the process *must* be planned. That means some of the lead time in the process is committed to via forecasted requirements. Inventory strategy should be mapped out, agreed to by both demand- and supply-side management, and rules of engagement documented. When the market practices or strategies change, high-performance organizations regularly update the handshake between the parties. *Again, the objective is lowest cost with highest service. To acknowledge the realities of inventory strategy is to allow for the most cost-effective and highest customer service processing of orders.*

Master scheduling different inventory strategies will make a difference in scheduling approach and what techniques will be used. That opens up the discussion on the definitions of product families.

PRODUCT FAMILY

Through lots of evolution within the ERP process, companies have determined that there is an optimum level of detail for planning at the top management level. This level of detail generically is referred to as product family. If product

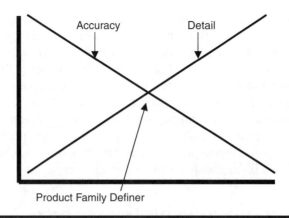

Figure 1.4. Accuracy Versus Detail Comparison

family has a specific meaning to you in your environment, do not get too involved yet trying to apply your definition. Product family can be defined differently in every business. The definition used in this book will be a common one, so set aside your definition for now. The good news is that you can call what in this book will be referred to as "product family" anything you want as long as everyone in your business understands it the same way.

As you can see from Figure 1.4, there is an optimum level of detail for the product family designation. This requires experimentation. It is difficult at best to try to define a generic formula to pinpoint the correct configuration that would work for all manufacturing companies. It just is not that simple. It is also common for a company to revise product families several times before settling on consistent groupings. Finance, forecasting, and operations planning all *must* share these common planning groupings. It is this insistence that allows the real power of the S&OP process to work and the master schedule to be most effective.

BILL OF MATERIAL

The SKU level to be scheduled by master scheduling is determined by the bill of material (BOM). The BOM is the recipe for how the product is put together and what components are required. The BOM records are one of the most important assets of an organization. They not only influence the manufacturing process but also impact the planning process in several ways. Generally, the more levels in the BOM, the greater the lead time and the more complicated the inventory strategy requirement will be.

BOMs do not have to be just the recipe for the finished product. They can be used for planning the unknown in master scheduling. Planning BOMs, as these special bills are called, are always used in high-performance operations. These specialized records may not always be called planning BOMs, but that is what they are. Planning BOMs drive components to be positioned in anticipation of a customer order. Good master scheduling always utilizes planning BOMs. Examples of their use might include engines and transmissions for auto manufacturing or resin for plastics molding. Complex subassemblies require signals to be in position prior to receipt of the real customer order. The use of planning BOMs will be discussed further in Chapter 5.

CAPACITY PLANNING IN THE MPS PROCESS

There are two types of capacity related to this discussion: demonstrated capacity and theoretical capacity. It is extremely important to distinguish between the two. In many businesses, highly energetic people with the best of intentions think that a carrot has to be out in front of the donkey's nose for the cart to be drawn at top speed. This may be true in lower performing plants, but not in high-performance ERP facilities.

At one facility, promises were made to corporate management that levels of production would reach new record volumes. Sales were surprisingly high, which was exactly what management wanted to hear. All it really did was buy time between the monthly reviews and create a false sense of security. The underpinnings were not there to actually carry out this promise. As will be demonstrated throughout this book, there is no place in master scheduling for overstated requirements. The most important job a master scheduler can do is keep the schedule tight and realistic. This is not possible without an understanding of demonstrated capacity. Capacity is a major concern for the master scheduler, maybe the most important factor governing the scheduling process. *A high-performance master scheduler always uses demonstrated capacity and is not easily fooled by opinions and emotions not backed by data or facts.*

Returning to our example, the master scheduler really suspected the truth about the capacity but went along with management. He did not feel it was his place to raise a red flag. He was very wrong! Additionally, unless demonstrated capacity is very consistent and repeatable, there is an argument to be made that some small amount of buffer time should be also considered. At Frigidaire's dishwasher plant in Kinston, North Carolina, this is called buffer "overspeed." Most high-performance facilities schedule some buffer time. No capacity is lost because if the schedule is on time, work is brought up from

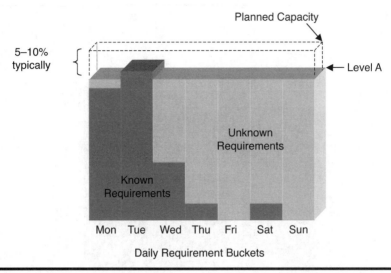

Figure 1.5. Master Schedule Known and Unknown Requirements

the next period. If this all sounds a bit confusing, it will be clarified as the story progresses.

In the previous depiction in Figure 1.2, "capacity" was shown at the top of the schedule. In Figure 1.5, level A is an imaginary level of capacity, just above the filled bucket value of capacity scheduled for consumption. The inference from the diagram is that the master scheduler should not schedule to the maximum demonstrated capacity all the time. To do so is to fail at master scheduling. Demonstrated capacity is what the process can generally yield. It does not mean that the process will always deliver at that level. Good master schedulers take their responsibilities very seriously. When they miss schedules, it is a reflection on their accuracy and performance. Class A master scheduling requires that schedules are hit 95+% of the time. To accomplish this, demonstrated capacity must be well understood and schedules continually synchronized with reality. In high-performance organizations, this must happen at least once a week. It is common for this to be done daily or even hourly, although if a good job is done initially, the frequency should be minimized. *In high-performance companies, synchronizing the schedule is not that difficult; often only small schedule tweaks are required. This practice should not be compromised.* It also requires recognition of known and unknown requirements (see Figure 1.6). Any consideration of an alternative practice is generally not in accordance with good master scheduling process.

While discussing reality, let's talk about customer orders. In most businesses, customer behavior generally is not predictable. Orders do not come in

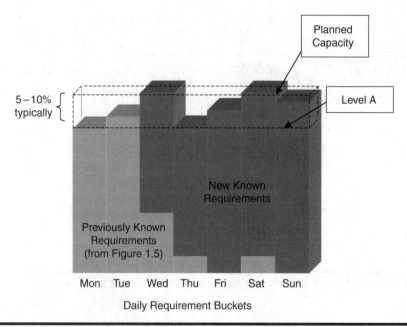

Figure 1.6.   Master Schedule Known and Unknown Requirements

at the same rate every day. Some days orders are heavy, and other days orders are light. If this sounds familiar, you are not alone. *Customers do not always work from a level schedule. They are reacting to their customers as well. Ultimately, it is the customers at the retail level that drive the whole supply chain.* Normal process variation might look like Figure 1.7.

In reality, this is probably how the orders come into your order entry process. No matter where you are in the food chain, consumer spending affects demand, and consumers can be finicky, to say the least. This is not how most

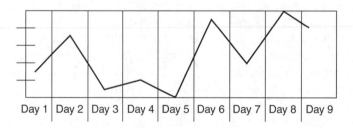

Figure 1.7.   Normal Customer Order Fluctuation

companies want to run their operation on the supply side. For that reason, some load smoothing is done regularly. This smoothing is initially done in the operations planning process, but the real schedule leveling takes place at the MPS. This is where planned orders meet firm customer orders.

MASSAGING THE DAILY SCHEDULE

When things are working well, the daily adjustments or tweaks to the schedule are minimal, but as most schedulers know, every day is not a "tweak" day! Sometimes unplanned activities happen, such as machines down, weather-related events, or even a bomb threat. In today's world, any number of possible surprises can affect the master schedule. There must be a process for dealing with these unplanned anomalies. That process is the daily schedule massaging done by most master schedulers. Figure 1.8 shows the results of a process with normal uneven demand complicated by a machine-down situation. If the original planned schedule was filled to the demonstrated capacity minus 5%, there would be some flexibility to move and shift orders without impacting customer promises in a big way.

This kind of constant massaging is necessary for many high-performance master schedulers. The root cause is customer variation, a process input that almost all businesses experience, regardless of market or product.

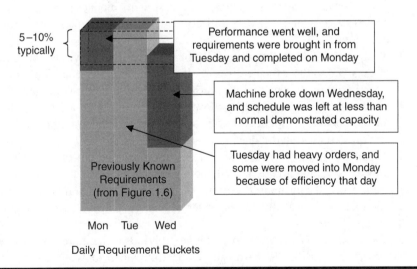

Figure 1.8. Massaging the Master Schedule Daily

RULES FOR MPS LEVEL LOADING

Level loading at the global level happens in the operations planning and will be dealt with in more detail in Chapter 3, but daily and/or hourly level loading is also necessary in most businesses. The following rules are appropriate after the operations monthly level loading has been accomplished:

1. Schedule materials to be available (as opposed to received) at least one period prior to the required need. If the scheduled time/requirement buckets are in days, allow supplier releases (only as authorized at this time) of material 24 hours ahead of requirements. If the schedule is in hour time buckets, schedule material to be available one hour prior to need. If the schedule is in minutes, schedule the material to be available one minute prior to need. (I have yet to see a schedule in seconds, but I can visualize it.) This may sound like waste, but if you want to execute to high levels of performance, material flexibility is required. Most businesses overdo it, however, by bringing material in way before the need. These businesses frequently perform at less than 15 to 20 turns. There is no reason in today's business environment to have less than 20 turns, and many high-performance businesses are getting more than 50. Only a few years ago, 50 turns was rare in manufacturing.

2. Schedule each product family separately. Along with the obvious reasons of shared constraints, this is necessary because inventory strategy affects product family selection, and this strategy dictates the level within the BOM at which the MPS will drive material, which facilitates proper scheduling.

3. Understand demonstrated capacity of the product family. Demonstrated means "normally executed levels," including normally experienced process variation (not including *unusual* process variation experienced).

4. Create weekly schedules that allow for some execution variation. Schedule some small amount of buffer time above scheduled completions up to full demonstrated capacity.

5. When performance is robust and no significant or out-of-the-ordinary process variation is experienced, move future requirements into the present period to fill any unused buffer. This can be done automatically from the factory floor if the internal rules of engagement are specific enough to this need.

6. If no orders exist to pull up in the next period or there are no customer orders convertible to cash, allow for employees to work on process improvement projects. Some suggestions include 5S projects (house-

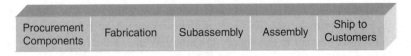

Figure 1.9. Accumulative Lead Time

keeping and workplace organization), setup reduction (both machine and assembly areas), preventative maintenance, etc. This is valuable time and must be put to good use.

FENCE RULES IN MASTER SCHEDULING

There are several theories and approaches to fence rules in scheduling. Class A rules of engagement normally include specifics for each product family in terms of time fence rules. Without these rules/handshakes, it is more difficult to have high performance with the optimum efficiency of operations. In a normal process, the lead time elements might look like Figure 1.9.

Your process may look different than this, but usually procurement exists and at least one other conversion process exists as raw material is converted into saleable product. If your process is simpler, you may have some interpretation to do. In Figure 1.10, the time fences are aligned to this lead time.

The "fixed" fence is sometimes referred to as the "frozen" fence. *No schedules are ever completely locked. Because of this, most companies have aban-*

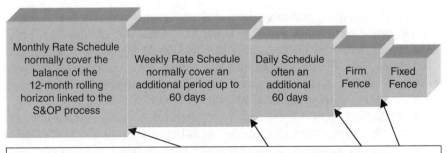

Figure 1.10.   Time Fence Norms and Flexibility Built into the Planning Horizon

doned the term "frozen." In Figure 1.10, the inference is that there is more flexibility built into the schedule. At each time fence, the rules change just a little and all the players in the supply chain are aware of the rules and the latest schedule. Flexibility is part of this handshake at each time fence break.

MANAGEMENT SYSTEM REQUIREMENTS

Management system is a term often associated with process governance. The term is used to describe an autopilot infrastructure for management follow-up. There are several management system events required in high-performance master scheduling processes. A main event that involves strictly the weekly MPS is the clear-to-build process. In this management system event, the master scheduler reviews the weekly schedule requirements for the upcoming week with the process owners that need to deliver them. Those players are the production managers and the procurement personnel. *The objective of the clear-to-build is to have clear visibility and accountability for the accuracy of each weekly plan. In high-performance organizations, this is a simple and quick process, but necessary nonetheless.* Since it is the objective of the master scheduler to have the highest level of accuracy possible in the MPS, it is without a doubt an important part of the master scheduling process. Management systems are covered in great detail in Chapter 10.

THE SALES AND OPERATIONS PLANNING PROCESS

The top management planning process in all high-performance organizations includes some form of an S&OP process. This discussion can best be started from a platform of ERP understanding, discussed at the beginning of this chapter (see Figure 1.1). Figure 1.11 depicts the normal ERP business system with master scheduling as the centerpiece. Top management planning is at the top of this business system, depicted by process boxes within the large circle.

The S&OP process is a management system for, by, and with top management to help manage the risk from the overall business process. *The S&OP is a handshake risk management process where both the demand- and supply-side managers gather once a month to review the likelihood that demand and operations plans will be met.* No high-performance business succeeds without some form of this process. It is known by many different names, including production, sales, and inventory (PSI) planning; sales, inventory, and operations planning (SIOP); and simply operations planning. Chapter 8 is devoted to this

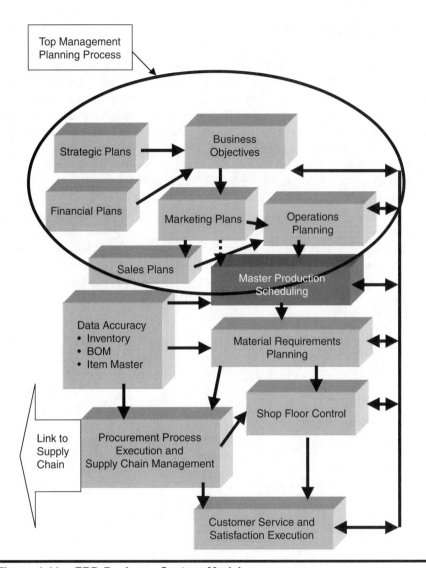

Figure 1.11.   ERP Business System Model

important process. The S&OP process is the major input to the master scheduling process, and without it, it is very difficult to effectively produce and execute an MPS. The master scheduler plays an important role in the S&OP process and often is the main player in prepping top management for the review.

PROCESS OWNERSHIP

Process ownership in the master scheduling environment includes making sure that the line schedules are in sync with the S&OP, accurate, and communicated properly. In addition, it means having a 12-month rolling horizon in the business system for the supply chain to see and maximize efficiencies. The schedules have to be updated/reviewed at least once a week in a Class A environment, and accuracy should be at least 95+%.

The process owner is obviously the master scheduler for this process. He or she will normally document the performance and opportunities and report to the weekly performance review to review accomplishments and progress with the operations team. The master scheduler also often facilitates the weekly clear-to-build.

The master schedule is the driver for all production activity, including the supply chain. It is a critical process and one of the most important in any business. The handoff from the MPS is the material planning functions within the organization flow chart. In the next chapter, inventory strategy, a prerequisite to the MPS, is discussed in greater detail.

This book has free materials available for download from the
Web Added Value™ Resource Center at www.jrosspub.com.

2

UTILIZING
INVENTORY STRATEGY

Inventory strategy was introduced in Chapter 1 and will be described in much more detail in this chapter. Inventory strategy is such a key factor that some businesses actually refer to this methodology as their *business* strategy. Dell, as an example, may fall into that category. Michael Dell, according to some texts, attributes the success of his computer business to the lean supply chain, access to customers, and not carrying inventory. This inventory strategy is not necessarily unique, but it was not common just a few years ago in the mature commodity markets. Most people would consider personal computers to be such a market. In Dell's case, its inventory strategy is assemble to order (ATO). In its planning process, Dell plans for the availability of components at one level down from final assembly. No final configurations are committed to until the customer actually calls with an order. Compare this to other more traditional computer manufacturing where stock might be held at a retail center, such as Best Buy, and another buffer of inventory is held in a warehouse somewhere for the retail draw. As you can see, the latter example, make to stock (MTS), is much more costly. Differences in customer service such as being able to touch and feel the product, some would argue, make a difference. I make no judgment as to customer perceptions on the topic of PCs, but the lessons are easy to understand and apply to any business. The business magic comes from picking the right strategy to give the most to your customer for the least cost.

In today's competitive world, just having the best product is not always enough. Most businesses are trying to move the inventory strategy up the food chain simply to reduce the amount of committed inventory required prior to

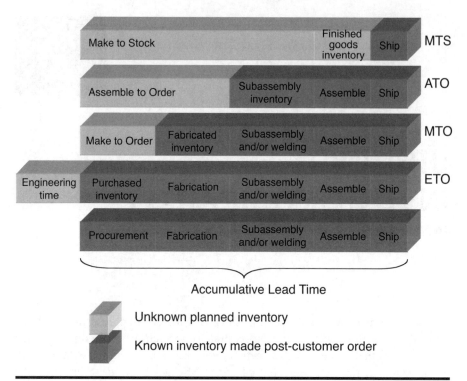

Figure 2.1. Inventory Strategy

receiving a customer order (see Figure 2.1). That means MTS companies are moving some product to ATO. Some ATO companies are moving more products to make to order. The reason is obvious. Less inventory required to run the business and still meet customer needs usually positively affects the cost of manufacturing.

The typical product life cycle is illustrated in Figure 2.2. Although some products do not follow this pattern exactly, many do. This is true even in the fast-moving consumer electronics business, where the inventory strategies link nicely to this product life cycle chart. In Figure 2.3, the strategies are added to the product life cycle model. New electronic revisions to existing products or new inventions normally start out as more expensive and more difficult to find and end up very common and low priced in the market. This is often true with capital goods, on the other side of the spectrum. Products that are not successful in finding a market do not complete the cycle. They tend to just fizzle out and die an early death, skipping the mature phase of the life cycle.

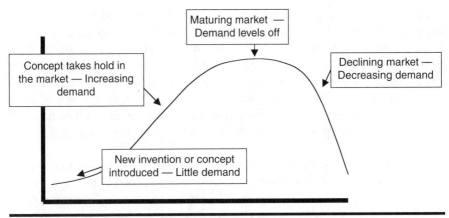

Figure 2.2.   Typical Product Life Cycle

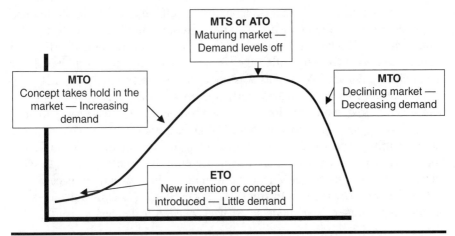

Figure 2.3.   Typical Product Life Cycle with Inventory Strategies

MAKE TO STOCK

MTS inventory strategy plagues most businesses and is a great place to start. The problems arise because the accumulative lead time is longer than the customer-acceptable lead time. This leaves no choice other than to either disappoint customers or stock inventory. When the answer is to stock finished goods, it is called MTS. The tempting side to this inventory strategy is that it provides ready products, which allows customer demand to be filled immediately. In some markets, this is essential to having any chance at competitive advantage. A company in Maryland that makes blow-molded bottles is in that category. Its only chance of keeping customers happy is to have inventory available when customers need it. A 16-ounce bullet bottle is available from many sources. Quality, service, and availability are the secrets to competitive advantage. I have been at this company's top management planning meetings and heard the vice president of sales voice his arguments to the side of more inventory. This can put the company into a risk management situation. When finished goods are generated and put into stock, both risks and benefits are increased.

Another major supplier in the packaging business that provides materials to fast-food companies all over North America made all products to MTS strategy as it started its journey to excellence. The problem in this particular market is that the product specifications can change any or every day. Finished goods are susceptible to obsolescence every day of the week. In this particular case, the company supplied products with graphics of sports figure or event or movie advertisements on them. When someone or something changed in the market, so did the graphics requirements. The only predictable thing was change. Inventory was good when things didn't change, but when something did, finished product was good for very little other than a donation to the local church for use at charity dinners. While this was good in some respects, it was not good for profitability. To make matters worse, there was more capacity in the market than was required, which put lots of pressure on price. The answer was to have less finished goods inventory without losing customer service. This required more flexibility, specifically the ability to change over product line configurations more quickly. Changes were required within the processes through setup reduction efforts. No master scheduling technique alone would be successful in adding flexibility without help from the process itself. Luckily for the stakeholders, management staff understood what needed to happen.

LEAN SETUP REDUCTION OR CHANGEOVER EFFECTS ON MTS STRATEGY

The most important step in getting organizations to change is sending strong signals to the employees that management has a plan and that it is consistent and tied to the management systems within the business. In the case of the fast-food company mentioned above, top management did the right thing by making setup reduction one of the biggest priorities in the business. At the time, the company was in the process of having all the plants meet a specific enterprise resource planning performance standard called Class A ERP. This status required meeting many different process performance levels and criteria, including top management goals established, communicated, and met. It was these goals that shook up the plant behavior. One of the first goals communicated to the multiplant organization was to "reduce overall setup time by 50% within 12 months." This goal alone would not be too earth shattering were it not for the second objective: "increase the number of setups by 50% within 12 months." This got the attention of the plant staff, even though in the beginning the plant managers in many cases were convinced that these goals were not possible.

In one plant in this company located near Boston, a changeover in a plastics operation regularly took over 24 hours each time it needed to be converted. A machine line in the process was in the range of 12 feet wide and 60 feet long. The various operations in this line were: extruding of plastic sheet, vacuum forming hundreds of parts in a multicavity mold, and blanking the parts from the sheets after forming and stacking the parts for packaging into corrugated boxes. Dies for this operation weighed thousands of pounds. Overhead cranes were used for hoisting these huge tools in and out of the process during changeover. It also should be noted that people who worked on this line were quick to tell how this changeover had been reduced several times over the years to the existing time standard of 24 hours.

The plant manager at this plant was fairly new to the position and did not have a predetermined opinion that the new company objectives of reducing setup time and increasing the number of setups were impossible. A team of workers from all shifts was assembled and charged with the project to reduce the setup by 75% (a huge increase over corporate goals). This sent a strong statement that plant management was serious. Plant management was sure that there was enough excess setup time to make this higher objective realistic. After only about 120 days of teamwork, using single minute exchange of die techniques, the result was setup time reduced to 1.6 hours (from 24!). This had an enormous impact on master scheduling, allowing a significant reduction in inventory.

SINGLE MINUTE EXCHANGE OF DIE

The most successful master scheduling processes are in combination with good lean manufacturing strategy. Since changeover time is waste, emphasis on less overall time is helpful on many fronts. Probably the most popular technique for minimizing setup time that has been very successful in literally thousands of businesses worldwide is single minute exchange of die (SMED). Developed by Toyota and credited to Dr. Shigeo Shingo, the SMED approach is an easily learned and communicated process methodology for many activities, including machine changeover and even preventative maintenance.

SMED starts with an understanding that all activity originates from two distinct processes: internal activity and external activity. To best explain the differences, think of a simple drill press like many hobbyist carpenters have in their shops. If the drill press operator is drilling three pieces of part A, which requires a half-inch hole, and then has to change to drill three pieces of part B, which requires a three-quarter-inch hole, there are some activities that would normally be planned and executed. The following list of activities may not be in the best sequence, but the activities need to happen in some order:

1. The half-inch drill bit is sharpened
2. The half-inch drill bit is brought to the machine
3. B parts are brought to the drill press
4. Completed A parts are moved on to the shipping department
5. The three-quarter-inch bit is removed from the drill press
6 The three-quarter-inch drill bit is sharpened for the next use
7. The three-quarter-inch bit is put away in its storage bin
8. The half-inch drill bit is inserted into the drill press

Using Dr. Shingo's methodology, each of these steps required by the process falls into either internal or external activity. Internal activity is activity that happens when the machine is shut off. External activity happens while the machine is still running. Any activity that can be external allows the machine to continue earning money. In the list of activities for the drill press, if the machine operator shut the machine down before doing any of the activities, they would be considered internal activities. In this methodology, internal activities are to be avoided or minimized if at all possible to keep the machine running. Upon further examination, it appears that the only steps that really require the machine to be stopped are steps 5 and 8. All the others could be done with the machine still running, even if some help was required. This is the idea behind SMED. Make sure the machine is kept running but allow lots of changeovers.

Dr. Shingo used an approach that has been adopted by many firms around the world. His methodology started by establishing an objective: to reduce the changeover on a particular setup by 50%. Once this is successfully accomplished, the next assigned team will have the same goal: to reduce the setup by (another) 50%. By the time the third team is chartered to reduce an additional 50%, the setup is close to 10% of the original standard at the time the projects were started (see Figure 2.4). This approach is a good one that many companies have found to be helpful. This was the method used in the packaging plant near Boston discussed earlier.

By reducing the amount of changeover from 24 hours to 1.6 hours, the effect on master scheduling was enormous. This allowed many more changeovers to

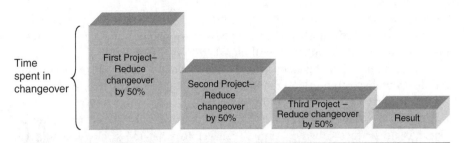

Figure 2.4. Cutting Changeover Time by 50%, Then 50%, Then 50%

happen in a week and reduced inventory significantly while actually improving customer service.

TAKT TIME EFFECTS ON MTS STRATEGY

Order size from the customer impacts the amount of stock any supplier has on hand. If, for example, the amount of material a customer orders is five truckloads each time material is ordered and there is usually only 24 hours from receipt of order to shipment, the manufacturing takt time must support the ability to respond if stock is not ready for shipment. If it takes more than 24 hours to produce five truckloads, some material will have to be in stock. This is normally the decision process that dictates an MTS strategy.

ASSEMBLE TO ORDER

Most organizations are trying to transition MTS business into ATO strategy. The reason is obvious. Inventory that is stored in the raw or subassembly state is much more flexible for use in customer-driven requirements. This becomes much more understandable when we talk about specific products. The first one that comes to mind was mentioned earlier — computers. If Dell used a business or inventory model that positioned inventory at the finished goods level, its expenses would be equal to that of competitors that use third-party distribution. The fact that Dell chose a different inventory strategy was important to its entire profit and service strategy. Whether Dell will be successful long term is still open for observation and debate, but the world has become more aware of the opportunities in recent years to lower inventory.

The usual way to successfully make this transition from MTS to ATO is to increase flexibility so as to allow the reduction of response time. This often happens as a reaction to improvement efforts where waste is an opportunity within the order fulfillment process. Processes are mapped and steps within the process eliminated. Suppliers are often asked to ready materials for reaction in short lead time. All of this adds up to lower cost and a wider range of configurations available, often giving competitive advantage. Efforts like lean and Six Sigma are driving many efforts in this space.

Inventory strategy is a strategic choice that every business has to make. There is no perfect "always correct" answer to what the strategy should be. There are examples of companies that have been very successful with every inventory strategy. With each strategy comes the need to combine it with the correct set of customer services and pricing. Some companies are even moving traditional MTS products to make to order. Each time a company successfully migrates its inventory strategy up the lead time line, it has the opportunity to reduce costs.

MAKE TO ORDER

Many companies are in the make-to-order (MTO) manufacturing space. These companies are able to turn raw materials into saleable product in time frames that equal the customer-desired lead time. Such companies include various commodities representing all types of products, from retail ice cream parlors to capital equipment manufacturers. Companies that choose to utilize MTO strategy plan and stock raw materials and convert them into saleable product in customer-acceptable time frames. Oceangoing ship manufacturers have customer-acceptable lead times that are reasonably long, allowing their strategy to revolve around MTO. Still other nontraditional MTO manufacturers are moving in this direction to save costs (see Figure 2.5). Many companies have special fast-track product configurations that can be made in very short lead time.

ENGINEER TO ORDER

Notice that within engineer to order (ETO), engineering time is actually planned rather than inventory. In essence, the engineering resource is the commodity being planned and "stocked," ready for engagement once a customer order is received. Any resource that is required for the manufacturing process must be planned or available within the lead time required by the customer. In the case of ETO product, in many markets there is not enough time to recruit, hire, train, and engage engineers after an order is received, although there are exceptions

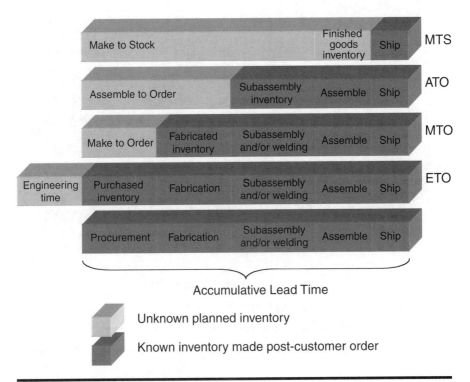

Figure 2.5.   Inventory Strategy

to that rule. Extra large capital goods, such as airplanes or submarines, can, in some cases, have longer customer-acceptable accumulative lead time that allows for ETO strategy to work. I have worked with several clients in that mode.

When getting deeply into master scheduling, there may be no more important topic than inventory strategy. It affects product family design, it affects what level in the bill of material the master production schedule (MPS) manages, it affects final assembly scheduling, it affects customer service, and it affects cash requirements in the business. As you read this chapter on inventory strategy, if you think you know what your business design is for inventory strategy, you may be wrong and will find inventory in places you were not expecting it. If at the end of your investigation you were right and the inventory strategy is as you thought, you deserve a pizza celebration!

Inventory strategy, also known in some circles as order fulfillment strategy or manufacturing strategy, is more about where you plan inventory than about filling orders. The MPS cannot be executed properly without that piece of

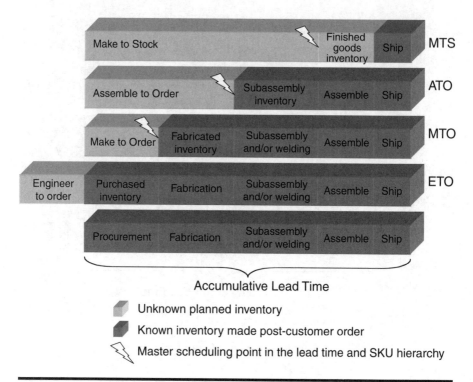

Figure 2.6.   Master Scheduling within Inventory Strategy

information. In fact, the MPS and inventory strategy are linked at the hip, so to speak. Figure 2.6 illustrates where the MPS intersects with the inventory strategy. It also introduces two more inventory strategies: ATO and ETO.

Say, for example, a company's bill of material structure looked like Figure 2.7. If the business determined its inventory strategy for assembly 123's product family to be MTS, the master schedule for that product family would drive requirements for the top-level parts in that family, in this case assembly 123. This means that if you went into the MPS to look at the specific part numbers scheduled, you would see the part number assembly 123 in the scheduled requirements going forward even though there was no specific customer order for the requirement. The plan would be driven from forecasted requirements. That is the characteristic of an MTS environment.

If the strategy was ATO, the subassemblies and/or weldments would be driven by the master schedule and available for the customer order when it was received. No final assembly would happen until the order was received from the customer. As depicted in Figure 2.6, sometimes only fabrications are driven by the MPS in an ATO environment.

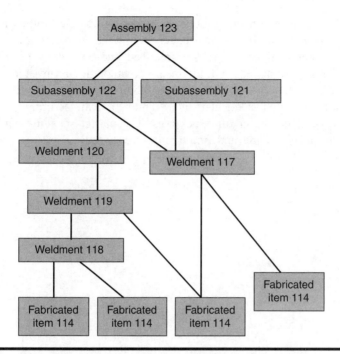

Figure 2.7.   Bill of Material Structure for MTS Assembly in Figure 2.6

In an MTO environment, the stockkeeping unit (SKU) driven by the master schedule might be the raw material SKUs or possibly the weldments, depending on the complexity and lead time of the customer requirement versus the accumulative lead time of operations. If you went into this shop at any time, you might find raw material stocked waiting for the customer order.

Each of these inventory strategies has its costs and benefits. Very few companies have only one strategy. Most have several. This is usually the most efficient way to run a manufacturing business, although there are always exceptions to that statement. Commodity products are most often the markets where products are made to stock. Learning in the lean space has recently been especially popular due to this extra cost. Commodity products are often the lowest margin products in the market, and the MTS strategy has more cost associated with it due to inventory requirements. Lean processes drive flexibility and speed, reducing inventory requirement and allowing more SKUs at the finished goods level to be moved to ATO strategy.

If an organization utilizes ETO, the MPS *could* master schedule the engineering time and resource. In one company I worked with that manufactured capital machinery for the automotive supply chain, the master scheduler planned capacity within the engineering group. Engineering was the limiting factor and the most critical constraint. By master scheduling the engineering resource, the right number of engineers would be retained and a strategy developed for subcontracting work as required. Engineering, although a professional need, is, after all, still just one of many requirements in the manufacturing process flow.

Inventory strategy is a key element of running a manufacturing business and is not always given the credit it deserves. The master schedule performance depends on understanding and fine-tuning this strategy.

This book has free materials available for download from the
Web Added Value™ Resource Center at www.jrosspub.com.

3

MASTER PRODUCTION SCHEDULE INTEGRATION WITH ENTERPRISE RESOURCE PLANNING

It has often been said that the most influential positions in a manufacturing operation are the plant manager and the master scheduler. That puts the master production scheduling job into perspective for some and makes the point that the master production schedule (MPS) is the score and the master scheduler the orchestra leader for the manufacturing symphony. To best start this next discussion, it is important to set the stage by examining the process within which the MPS resides in high-performance manufacturing environments. That process model is enterprise resource planning, more commonly known as ERP.

New "names" for ERP are invented every week by software companies or consulting firms trying to sell their wares. Some new acronyms stick and some do not. Being in the business of sorting this stuff out, I prefer to stick to the ideas I have seen add value. In the years I have been involved with process improvement, I have determined that some of these process methodologies are, in fact, legitimate. Three specifically are: (1) Class A ERP, (2) lean, and (3) Six Sigma. There are some specific goodness factors embedded in each of these approaches, and an argument can even be made that although these process models are best done concurrently, the initial focus might best be done sequentially, even if quite quickly. Because the MPS is such a major component of the ERP business model and especially important within Class A ERP performance, the emphasis at the start of this discussion is appropriately focused on that process. Lean thinking is also especially important, but not efficiently until there is a solid foundation of discipline and process repeatability. Developing traction is much easier once the culture is tuned to accountability with management systems in place. This is the essence of the ERP process done well.

There is a lot of confusion about the topic of ERP. This confusion is often credited to the aggressiveness of business system software companies. Not that ERP tools do not offer a lot of value add, because they do. Tiger Woods would have had a difficult time maintaining a high ranking in the world of golf if he did not have quality, straight and true golf clubs and perfectly balanced golf balls to limit the process variation. The rest of the story is meaningful as well, however. Without his skillful and disciplined execution, the tools alone would not get the job done to a high-performance standard.

Like a professional-level game of golf, high-performance manufacturing requires disciplines and execution with the highest process repeatability and predictability. ERP is a popular business model that involves all levels of the organization. ERP process disciplines allow organizations to link top management decisions all the way through to execution in the supply chain and the factory floor. Well-executed ERP not only starts with top management but is totally dependent on top management. The MPS is at the heart of this business model.

TOP MANAGEMENT PLANNING

Top management planning is clearly the most important process within ERP. This planning process is the enabler for the MPS process among other processes in the business that take their lead from management decisions. As illustrated in Figure 3.1, it is clear that many decisions must be made to correctly guide the organization's vision, including the MPS. For the ERP process to work well and the MPS to be effective, there needs to be a direct connection between management direction and execution. ERP is the nervous system and information flow that link top management thinking and planning with marketing, sales, capacity planning, procurement, manufacturing, and customer service. The top management planning engine is worthless if there is no transmission to directly link top management thinking with the execution of those plans. When all the pieces are in place, the MPS plays that role.

Class A performance within an ERP business model applies accountability and management systems for sustainability. Management systems, in a manufacturing business, are planned and executed accountability infrastructures that

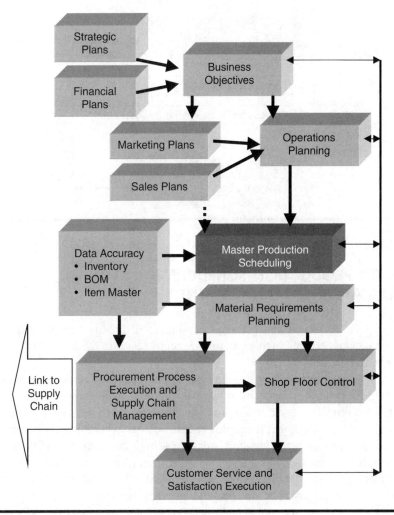

Figure 3.1. ERP Business System Model

create predictable opportunities for follow-up on decisions and goals. Examples would include infrastructure events such as daily performance reviews, weekly project management reviews, top management monthly planning performance reviews (often referred to as sales and operations planning reviews), and events directly related to master scheduling, such as a "clear-to-build" process. The clear-to-build infrastructure will be described in much greater detail later in this text.

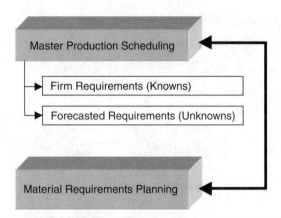

Figure 3.2.   Master Scheduling

MASTER PRODUCTION SCHEDULING
AND MATERIALS MANAGEMENT

The outputs from top management planning feed directly into the MPS and materials management when executing correctly. Figure 3.2 illustrates the subsequent activities.

While master scheduling in many businesses is a clerical process of documenting the promises made by order management, when done properly it is much, much more. Master scheduling, usually referred to as master production scheduling, is anything but a simple process. Master scheduling done correctly is a science *and* an art. The short story on a high-performance MPS process starts with the idea that top management planning, known as sales and operations planning (S&OP), drives product family demand and operations expectations in both supply-side and demand-side requirements.

S&OP (or sales, inventory, and operations planning or SIOP as it is called in some organizations) has been recognized as a critical process for years, maybe as many as 25. In recent years, it has been especially well recognized within high-performance circles and has gotten a lot of press. Done correctly, the S&OP process outputs are stated as product family requirements. This process is the handoff to the MPS. These product family signals are directly translated into detailed firm and forecasted signals by the MPS. When the ERP process works efficiently and effectively, scheduling becomes the linkage rod or transmission that connects the top management planning engine with the execution wing of the business, operations. When the ERP process is not working properly, the linkage often does not exist, or if it does, it is weak and loses fidelity through to execution of plans. The scheduling process often becomes a stand-alone function reacting to daily demand with short planning horizons. The results can range from ineffectiveness to devastation, and rarely is the outcome long-term success. Most of today's best-managed businesses hold the S&OP process as a valuable component of their top-level management system. As Tim Frank, CEO of Grafco PET Packaging Corporation, said to me, "Why wouldn't any top manager want to do this?"

Master scheduling takes these signals from top management and translates them into usable requirements for production planning. It is the master scheduler's job to determine the mix within product families and other unknowns, such as lot size and priority of order requirements. Without these important communication links, material requirements planning (MRP), the ERP planning engine, would not have anything to plan, or if it did, no completion requirements would be posted other than actual orders. Much of what the master schedule sequences are "unknowns" (see Figure 3.3).

Unknowns are often the drivers of unnecessary inventory buffer, so it is critically important to take this scheduling process seriously. The best master schedulers work with their organization's supply chain managers, both internal and external, to minimize lead time and increase flexibility within the supply base and at the same time work with the order management team to get fast rules of engagement that put everybody on common ground in terms of goals and expectations. The master scheduler's job is much like the conductor in an

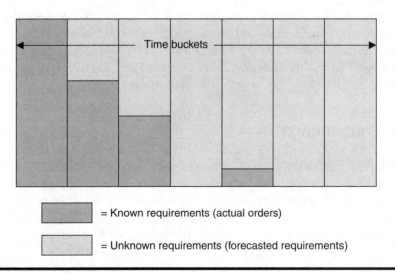

Figure 3.3. Scheduling "Knowns" and "Unknowns"

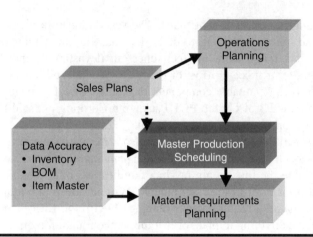

Figure 3.4. Data Accuracy within ERP

orchestra. The score documents customer need, and the master scheduler interprets the musical score (customer need) by adjusting the tempo or making slight changes as appropriate. There are many players in manufacturing and supply chain management, just as there are in an orchestra, and they all need to be coordinated. The baton's drumbeat is the master schedule itself, broadcast to the team. When everyone is exactly in sync, at the end of the song there is no extra unplanned inventory buffer and the music was played exactly to the customer's liking.

There are several inputs to the MPS, and the more accurate these data are, the more likely the process will be successful and value add. For this reason, high-performance or "Class A" ERP performance also includes data accuracy as a specific area of focus. Figure 3.4 is a segment from the ERP business model showing the interaction of some of the data accuracy elements.

DATA ACCURACY

ERP inputs play an important role in effectiveness of the overall ERP process. Accuracy of the inputs obviously plays a critical role. Data integrity is a historically proven prerequisite to high performance. It has been an asset for process improvement as well as process predictability in all high-performance organizations. In an ERP environment, data accuracy focus elements generally include inventory balance accuracy, bills of material accuracy, and other item master file field data accuracy elements, such as routings, lead time, and cost standards. Since the ERP process focuses on data and process linkage, it be-

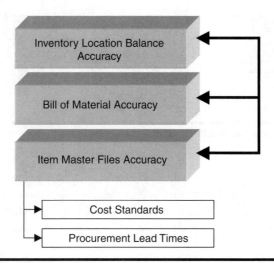

Figure 3.5.   Database Accuracy

comes extremely important for the data elements to have high integrity. *When inputs to the ERP process such as inventory balances (which directly affect MRP) are accurate, the user has a significantly improved chance of producing valuable and accurate outputs such as supply chain signals and shop floor orders.* Examples of these accuracy-dependent processes include not only requirements planning but also cost planning and inventory management (see Figure 3.5).

BILL OF MATERIAL ACCURACY

Bill of material (BOM) accuracy is possibly of even more importance than inventory balance integrity (although no one should argue that either is unimportant). BOMs are the recipes that are used by many organizational functions. This would include kits in distribution and assemblies, batch mixes, weldments, and kits in manufacturing. If this information is not right, it becomes difficult to cost-effectively deliver a quality product to the customer. If wrong, the BOM either has wrong components, and informal processes to get the correct materials in place, specifications are wrong or product is reworked before it leaves the building. Regardless of what is wrong, it is unnecessary process variation and cost. All high-performance organizations recognize the importance of this information and treat it accordingly. Few would argue against the value-add characteristic of accuracy in the building of requirements for engineering de-

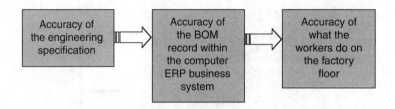

Figure 3.6.   Three Elements of BOM Accuracy

sign, material requirements, quality standards, and cost planning. Not only is it important that the BOM record within the ERP business system meet the engineering specification, but these records should also match what the factory floor actually does! There are three areas that need to be matched (see Figure 3.6).

INVENTORY BALANCE ACCURACY

Inventory accuracy is another key element of ERP data management. Inventory records are the jewels of the planning process and affect both requirements planning and costs. In the planning process, there are two types of inventory records: on hand and on order. Both need to be accurate for the planning process to work properly. Like so many other topics covered in this ERP overview chapter, inventory accuracy will be given ample attention later in the book as we get into each element of the MPS in detail. Figure 3.7 illustrates the interaction of data records with the planning engine of ERP, master scheduling and MRP.

If the inputs are not absolutely accurate in this model, the outputs directing shop floor activity and supply chain management are corrupted and therefore also not necessarily accurate. Too many organizations do not realize the process variation costs incurred by ignoring this fact. *The more accurate the inputs to the planning process, the more likely the outcomes are to be accurate and constructive toward the goals of the organization.* Given the fact that demand forecast accuracy will never be perfect, it behooves all manufacturing organizations to have other controlled data as perfect as possible.

EXECUTION PROCESSES

The execution processes of ERP are also outputs of the planning process and are literally the end of the trail where ERP outputs from plans at all levels are

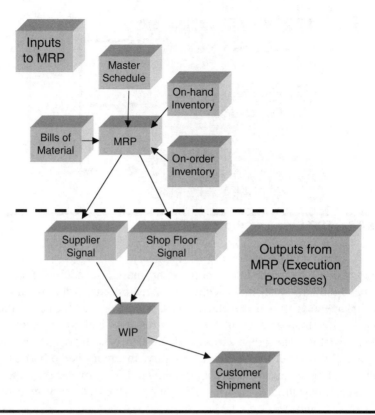

Figure 3.7.   MRP Inputs and Outputs

executed (see outputs in Figure 3.7). These include procurement, shop floor controls, product delivery, and service execution. Supply chain management is done from this vantage point. Without these signals, the supply chain cannot be linked solidly to the manufacturer's vision and drumbeat. *In many manufacturing organizations, purchasing functions under confident but less informed management act quite independently and do not have tight linkage to the MPS process.* In high-performance ERP organizations, procurement processes are tied directly to the ERP planning processes just as tightly as other ERP outputs. The benefits of this linkage discipline become obvious. The importance of the MPS role in this linkage will become more and more obvious as the discussion unfolds.

Inventory is obviously a direct result of purchases. Buying the wrong inventory means wasted energy and money on several levels. Buying too much inventory is also wasteful in a cost-conscious competitive world. Only disciplined process linkages to the information flow initiated by the top management

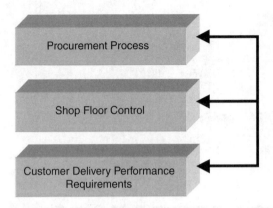

Figure 3.8. ERP Execution Processes

planning process and scheduled through the master scheduling process can allow maximum efficiency and accuracy of inventory-buy commitments. Without the linkage, inventory is driven from best pricing or truckload quantities and not necessarily strategic amounts. This *can* result in additional obsolescence, stranded inventory, and other unnecessary cost drivers. Inventory is often referred to as an evil, but it is quite the contrary. Inventory is the best asset you can have at the time the customer comes to buy it. ERP provides the opportunity for this synchronization to happen. Discipline is the necessary glue to keep process linkage intact.

The MPS is the map and driver for the execution processes (see Figure 3.8). Without this direction, the shop and suppliers would not have clear directives and chaos would result. In organizations where disciplines are lacking in the MPS to execution processes, inventory is often higher than necessary, service is less than effective, and costs from expediting are often high.

ERP AS THE STARTING POINT

ERP, as briefly explained in this chapter, is the overall business model that defines information flow and accountability. In their simplest form, high-performance applications of ERP processes are repeatable and predictable. The focus is generally on data and schedule accuracy.

Lean, as a separate improvement focus topic, would normally not reference a business model, as would a discussion of ERP. Lean is a different element. *Whereas ERP is a process, lean is more of an approach within a process.* For that reason, lean lends itself well to be a separate but integrated focus on the journey to excellence. Lean refers to an improvement approach that is focused on waste elimination. Lean thinking in a business is about looking at all processes, even repeatable processes, as opportunities for cost reduction and customer service improvement. This is normally done through rigorous process evaluations using mapping and other problem-solving tools.

At this point in the discussion, ERP emphasis has been on process and information flow predictability, followed by a lean focus on improving the same processes through the elimination of waste. Nothing new here for most readers, but you may begin to get a sense of the efficiency gained through the sequence of focus. That is exactly the same sense most successful businesses all over the world have determined. The issues are all the same. Quickly get the processes to work repeatably and then come back to improve the processes through a refreshed lean focus.

So where does Six Sigma come into this discussion? When reading about Six Sigma and relating back to their statistical process control experiences of the 1980s, many managers have thought: "How the heck will complex statistics with a goal of zero defects help real-world problems on the shop floor? Those Six Sigma people live in a dream world."

The best approach starts with people understanding the basics. Call it what you want, but these basics go back to the ERP discussion. Six Sigma is a sophisticated vision built on an organization of problem-solving and process-improving teams and tools. Once an organization is ready for this level of cultural focus for tenacious improvement, Six Sigma is the right methodology. The reality is that it takes organizational maturity for this to work. Many *successful* organizations have taken the improvement process in progressive steps: (1) ERP basics, (2) lean thinking, and mature to (3) Six Sigma quality.

Six Sigma is a powerful set of tools and a methodology that can empower a disciplined and trained organization to go from good performance to great. Six Sigma in its most literal sense is a designation of quality defined by standard deviations and defects.

As illustrated in Figure 3.9, when you look at the specialty and strength of each of these improvement methods, you find they complement each other quite well. In almost every company I have been involved with, I have found this to be applicable and a valuable understanding to start with. Having a good set of ERP standard practices that predictably result in promises met is a great foundation upon which to build a culture of continuous improvement. The logical next step in many organizations is not to jump right into complex statistical tools and deal with the hours and hours of in-depth training required to establish Six Sigma green belts and black belts. While there is a time and place for that, it is more efficient to keep the process focused on significant yet incremental improvements with a clear vision of where you are headed, includ-

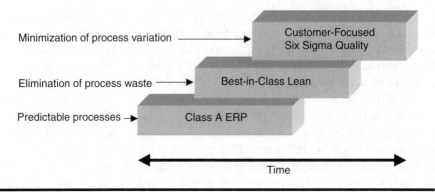

Figure 3.9. World Class Steps

ing the expectation of Six Sigma levels of quality and process reliability (see Figure 3.10). Be acutely aware that this is not to downplay the importance of Six Sigma process methodology; quite the contrary. *Six Sigma is a very powerful tool, and when organizations, including management, are ready for this level of focus, it is the right step to take.*

The final step in the journey is world class performance. There will probably be more steps to come in the future. Let's hope we live long enough to see them, but it is difficult to see beyond world class performance. In fact, I am a little foggy on that level. In many well-schooled experts' minds, world class performance is not yet clearly defined. It also changes every day, and to make matters even more confusing, world class performance is not defined by the same

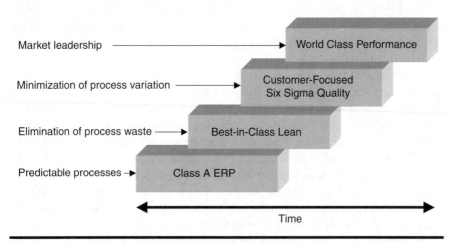

Figure 3.10. Journey to World Class Performance

market week to week or process to process. It is probable that technology will continue to play an expanding role in business management, and that last step may well be governed by techniques of technology use. Who knows? We can only imagine the world in 20 years.

Class A ERP performance is a major step in the right direction in every business that focuses on it, and the management system requirements from Class A ERP greatly help an organization mature in terms of both accountability and discipline. It is the foundation upon which to build strong manufacturing process. The MPS is a integral part of this process. Without good process in the scheduling and planning linkage to execution, high performance simply is not possible.

This book has free materials available for download from the
Web Added Value™ Resource Center at www.jrosspub.com.

4

DEMAND PLANNING INFLUENCES

Forecasting is probably the single most talked about process within the enterprise resource planning (ERP) business model. Master schedulers are involved with the demand plan forecast so frequently and deeply that they often feel ownership in it. Unfortunately, in many businesses the demand plan or forecast is *actually done* by the operations people because the demand side has not yet willingly come to the planning table to give forecasting a decent chance. *In high-performance business environments, a solid handshake is established between the demand-side process and the supply side of the business. Accountability is obvious and everybody is on the same team.* This handshake culminates in the sales and operations planning process. This partnership attitude is important not only for minimizing risk, but it also creates an environment where everyone has shared goals and understands how variation from their part of the business affects other parts of the operation.

With the demand plan as a major element of the top management planning process, it becomes imperative that the master scheduler be close to both the demand plan and the process used to develop it. While the sales and marketing people have the closest knowledge, the operations people have good feedback as well. Past realities are a good input to the process, but certainly not the only or highest value-add element. The inputs to the demand plan come from many areas and should all be taken seriously in organizations with expectations of improvement. Figure 4.1 illustrates most of the more important inputs. Keep in mind that economic factors are inputs to the business plan and are included in that process prior to the demand planning process.

The inputs are what make the quality of the demand plan. If the emphasis is on the right objectives and processes and they are linked to the business plan

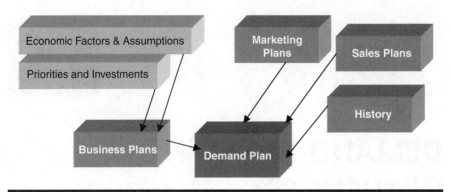

Figure 4.1.   ERP Demand Plan Inputs

properly, the results are usually quite fruitful. Demand planning effectiveness does not stop with the development of the demand plan. This plan has to be well communicated, well understood by the management team at all levels, and the objectives shared. *In too many organizations, the forecast is the excuse for all process variation within the operation. It just is not that simple. Any organization that requires an accurate forecast to excel in its marketplace is doomed before it starts!*

Instead of talking only about the weakness in the forecasting, the emphasis also needs to be on the lack of flexibility in the operations plan. The master scheduler plays a pivotal role in this leadership. In many organizations, the master scheduler leads the pack on awareness of operational requirements to ready the organization for the short lead times and quick response required by most markets today. While master schedulers may not be able to single-handedly change the culture of their organizations, their influence is normally well respected and their leadership can make great strides working in lockstep with the plant manager or other leader.

Keeping in the spirit of demand planning, understanding the background and inputs to the forecast can be helpful in putting the process requirements into perspective (see Figure 4.2). The logical place to start the discussion is with the most important and influential input, the business plan.

BUSINESS PLANNING AS AN INPUT TO DEMAND PLANNING

In businesses that are continually developing competitive advantage, it is important for the plans and actions to be tightly linked. As management direction

Figure 4.2.   Class A ERP Demand Planning Inputs

influences new product development, new markets, and service offerings, it is extremely helpful if everyone is on the same sheet of music. The business plan inputs are probably the most important to a company's success. It is this input that directs specific resource assignment in marketing and sales and redirects goals. It is not just about following the customers' needs; it is about influencing or leading them to your strengths whenever possible. This is what frequently gets lost in the forecasting process.

High-performance businesses are focused on getting their strategies implemented. Most companies are trying to increase their sales and grow the business. There are only two ways to grow a business:

1. Get new customers
2. Get existing customers to buy more

This is management's job. These are the strategies that management should be focused on. Who are the new customers and what are the additional demands coming from existing customers? Without a plan to achieve the two requirements listed above, the plan will not happen without divine intervention or pure luck, and luck usually has its limits.

When companies understand their markets and lead their customers to their strengths, the result is usually satisfied customers and successful transactions. The master scheduler role includes being cognizant of these strategies so that resources are planned to meet anticipated demand. The forecast alone does not always shine a bright enough headlight, especially given the dangers and likely potential for inaccuracy. Sometimes new products can take off unexpectedly, or worse, the opposite can happen. When left to chance, the result is a crapshoot as to readiness. *Operations managers (including master schedulers) who blame their own ineffectiveness on forecast inaccuracies are too often not stepping up to the requirements of the market.* If your market is not predictable and new programs, products, and/or services are being offered that cause process variation, the realities become obvious. In such cases, flexibility and responsiveness need to be an integral part of the scheduling equation. Lean strategies weaved together with good inventory strategy can help significantly.

MARKETING PLANS AS AN INPUT TO DEMAND PLANNING

When operations people think about forecasts, they often think about *guessing* what customers are going to do in terms of product demand. It is common to hear statements like, "If we could just know ahead of time what customers were going to want, we could run very effectively and with very little inventory." This kind of thinking is only half right. In high-performance businesses, the correct thought process for demand planning is not to *guess* what customers are going to do but, instead, *affect* what they are going to do (see Figure 4.3). Most businesses today are not trying to repeat yesterday; they are trying to do things differently to continuously improve position in the marketplace. This means that marketing needs to estimate what impact each of its strategies will have on customer behavior. That requires commitment from the marketing team. In a well-managed ERP environment, this accountability is referred to as process ownership.

For example, suppose the business plan outlines two new products for design and introduction. One is in a totally new market, and the other is a new design in an existing market (which addresses the two necessary elements for business growth listed earlier). Marketing will usually align with the new product introduction teams and have people assigned to each focus area. If the new product

Figure 4.3.   Class A ERP Demand Planning Inputs

that is also a new market for this imaginary company has three marketing ideas to implement, marketing needs to link its plan to some sort of expectation.

As an old boss of mine used to say, "There's always math behind everything." Let's get to the math behind this company strategy. Suppose the three ideas include one trade show, one customer event, and one advertising campaign. If the marketing group puts a value on each, some effect would be calculable on expected demand. Let's use an estimate of 30,000 units if nothing is done except introduce the new product through existing markets. Splash from the planned trade show might add another 10,000 units. The estimate is best done if the potential new customers are listed with anticipated buy quantities. The customer event designed to create excitement for a few of the biggest potential customers for this product could add another 50,000 units. Lastly, the advertising campaign might have an effect of 25,000 incremental units. Doing the math, the marketing team would add 30,000 (preexisting potential) plus 10,000 (trade show), plus 50,000 (customer event), plus 25,000 (advertising campaign), for a total of 115,000. If this math became the forecast, it would be a miracle if the 115,000 total was accurate. Marketing needs to estimate process variation just like the operations people have to. As uncomfortable as it may be for the operations people to hear, it needs to be said that it is also probably better to be a little off on the low side than to have way too much inventory and nobody buying it. Marketing needs to factor the estimates. What makes this process powerful is the process ownership. The factoring has some risks, and the marketing people need to know they will be measured on their accuracy. Hopefully, the master scheduler is fully cognizant of the risks as well as the forecast. Good schedulers do not run blindly on the forecast and point fingers when it fails them. The two sides, operations and sales/marketing, must have a handshake on a risk management plan.

PRODUCT PLANNING IN CLASS A ERP

Specifics on how the master scheduler would deal with the forecast will be provided later, but now is the right time to introduce the concept of product planning into the discussion. This topic could be a major key to successful market competitiveness and share growth. Good product planning in a business establishes market drivers and assumptions around which the rest of the business can build process. What would make sense for the level of detail expected in the forecast? The answer has been proven time and time again and is a critical point for clarity.

Demand accuracy is obviously most helpful if it is accurate. Nothing new there. It is also just as obvious that high accuracy is not likely sustainable over

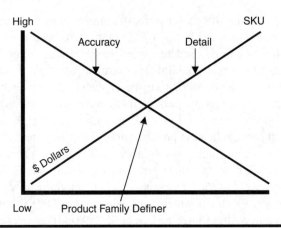

Figure 4.4. Accuracy Versus Detail Comparison (from Chapter 3)

time, especially at high levels of detail. It is for that reason that finding the right level of detail is important, not only for the sake of accuracy but also for planning the right inventory strategy. The best master scheduling processes insist that the business planning, demand planning, and operations planning processes all be in the same denominations. See Figure 4.4 to visualize the relationship between accuracy and detail. The lowest level of detail might be dollars of revenue forecasted. The highest might be stockkeeping units.

At the product family level, companies need to see the highest level of detail at the highest level of accuracy. This optimum level of detail takes some experimenting. If both marketing and operations have the shared goal of accuracy in this metric, the job can actually be fun. The following rules work best. Writing them on the wall in the main conference room could be helpful. If an organization can get this agreement between the two groups, mountains can be moved.

1. Operations should not point fingers at marketing for missing forecasts. Instead, operations should agree to help marketing succeed. Forecasting accuracy is much more difficult than accurately scheduling manufacturing.
2. Marketing needs to step up to the bar and understand and acknowledge that the farther the forecast is off, the more cost potential is generated for the company's products.
3. When the forecast is off substantially, the sales organization needs to help with the damage control and not just expect manufacturing to "jump through hoops."

Figure 4.5.   Class A ERP Demand Planning Inputs

SALES PLANNING AS AN INPUT TO DEMAND PLANNING

The next input area to look at is the sales process (see Figure 4.5). This plan is more calculable than the marketing plans. In many businesses, there is experience to work from in terms of closure rates and sales cycle times. In capital equipment, for example, there is a normally a relatively long sales cycle, but it is known and can be reasonably predictable over a broad customer base. Other consumer products can have no time at all. In that case, customer reaction is also relatively predictable. Sales also are often linked to seasonality or some sort of cyclicality.

Getting the "math behind the process" in the sales function is critical. This can mean predictability in number of customer calls, frequency of visits, follow-up on late orders, number of cold calls, etc. Remember that although the business plan (which states the focus of changes in sales volume and where the focus will be) is a major driver in the measurements that could be appropriate in your business, the sales plan is close behind. *The job of the sales organization is not to just guess what customers are going to do; it is their job to affect customer behavior.*

HISTORICAL INPUTS TO THE DEMAND PLANNING PROCESS

The last input to the demand planning process is the one that is, unfortunately, most often thought of first, historical data (see Figure 4.6). The reason it is thought of first is most likely because, in many businesses, it has been the role of operations to do forecasting. Since the operations people know very little

Figure 4.6.   Class A ERP Demand Planning Inputs

about what is happening in the marketplace (they are not out in it), the only data they can use that helps is historical. In many businesses, history projected forward is interestingly accurate, giving a false sense of security. That is not always good. Businesses that can accurately predict the future using the past often are not changing the landscape in what for many are changing markets. This can often come back to haunt an organization in the form of lost market share. This is not to say history is not valuable. Quite the contrary. Within the data are the seasonal stories and the normal cyclicality that are so necessary to understand.

Today, there are also many good software tools that can organize data and historical information to help statistically predict future events. They are helpful in organizing customer data and applying the data to the master production schedule. These tools have come a long way in the last few years and should be considered if your ERP business system does not handle these activities proficiently.

History is the most difficult input to use accurately for demand planning in businesses that are making waves in their markets. If you are shaking things up with new products and services, history becomes less likely to be accurate or even as valuable. Historical data can also be manipulated in many ways. Software forecasting equations that allow weighting in any period you choose actually allow you to get whatever result you want. That is both good and bad. *The bottom line — keep history in perspective and use it wisely!*

UNDERSTANDING THE OUTPUTS OF DEMAND PLANNING

The demand planning process is done to help a business understand profit potential. It sets the stage for capacity, financing, and stakeholder confidence.

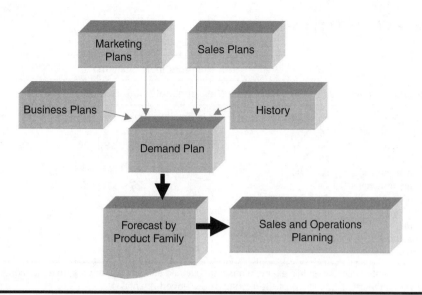

Figure 4.7.   Class A ERP Demand Planning Outputs

The forecast is probably the most valuable input to the risk management of a business. At the sales and operations planning (S&OP) process meeting, top management will evaluate the likelihood of the demand plan being accurately executed and accordingly will commit resources and cash. Demand planning is not one-way communication or a plan that is thrown over the wall to manufacturing. Instead, it is a handshake agreement between all parties. In essence, everyone tries to keep everyone else honest. This system of checks and balances helps with the risk management process.

The output of the demand plan is the forecast (see Figure 4.7). It should be divided into product families and have a horizon of 12 rolling months. This is necessary to meet the requirements of the S&OP process (see Table 4.1). The S&OP process will be discussed in greater detail in Chapter 8.

THE WEEKLY DEMAND REVIEW

The demand plan is a moving document due to the simple fact that customers are generally ill-behaved. Each week companies get smarter than just the week before. It is foolish to ignore that fact. For that reason, a demand review should be held each Friday with all product managers, plant managers, and master scheduler(s). In most businesses, this allows changes in the current week to be

Table 4.1.   S&OP Process Spreadsheet

Date: 2 February 200_

Product Family	(All Numbers in Thousands)												
	Jan*	Feb	Mar	Apr	May	Jun	Jul	Aug	Sep	Oct	Nov	Dec	Jan
Line one	30	30	20	25	34	35	45	78	67	50	35	35	30
Actual	30												
Line two	30	27	22	21	20	25	15	20	19	16	15	12	10
Actual	29												
Line three existing	60	80	86	90	88	89	90	78	65	50	35	30	32
Line three new**								20	25	50	75	80	90
Actual	66												
Line four	50	51	55	65	64	66	73	78	67	55	56	55	50
Actual	54												
Line five	5	8	6	7	7	7	8	9	10	9	7	6	6
Actual	5												

* The January column in this example has both planned and actual numbers from the month's performance; the balance of the months are planned quantities.
** New products are generally reviewed separately from existing products due to the high-risk opportunity.

quickly implemented and baked into the schedule changes for the upcoming week. In some businesses, this may mean changes to the supplier signals (make to order or engineer to order), in others changes in the manufacturing schedule (make to order or assemble to order), and in still others changes in distribution (make to stock). It depends on your business's inventory strategy — whether you keep your inventory in finished goods, in components, or at the supplier's facility.

The master scheduler is the coordinator or facilitator of the weekly demand review in many companies. The demand review normally happens at the same time each week with the same players. This way, no one can say they did not get the e-mail or receive notice of the meeting. A standard format should also be followed to ensure the right things are discussed. Using Lotus Notes or Windows Outlook, schedule the meeting as an ongoing, reoccurring meeting. This way, it automatically shows up on the calendar of everyone who should attend. The meeting can be held in a central meeting room or it can be conducted by phone. Conferences calls are often the only option for larger companies where logistics make it impossible or too costly to meet in a common location.

At one roofing company with headquarters and sales located in Massachu-setts and manufacturing in North Carolina, the weekly demand review is con-ducted using a videoconferencing setup each Friday. Each week, management from the factory meets "face to face" via camera and television monitor with management from the demand side hundreds of miles away. It works well, and the addition of video allows improved communication, not always evident in phone-only remote conversations.

Agenda for the Weekly Demand Review

The agenda should remain stable, with the master scheduler leading the charge with operations data for each topic. The demand manager will often bring the demand-side data for comparison and analysis. The agenda should include the following items:

1. Review current week's accuracy of forecast for each product family
2. Review upcoming week's forecast and any changes necessary
3. Agree on any required adjustments to the production plan
4. Determine any effect on the month's revenue or profit
5. Review actions and agreements from last meeting and new ones for current week

Attendees at the Weekly Demand Review

The attendees are somewhat different from one business to another due to size and structure as well as job titles, but the general expectation requires process owners from the demand plan to be on the call or at the meeting. A normal list of attendees might include:

1. Master scheduler(s) (supply side)
2. Product managers (demand side)
3. Plant managers (supply side)
4. Operations vice president (supply side)
5. Optional attendees
   a. Sales vice president (demand side)
   b. Marketing vice president (demand side)

In some larger organizations, people come in and out of the meeting according to prescheduled time slots for specific product families. For example, the product manager for line 2 does not need to be on the call when line 1 is being discussed unless it is of some specific interest due to shared resources, etc. If multiplants are involved, usually master schedulers at each facility are only interested in lines that affect their facility and schedule. This discussion is not

Table 4.2. Monthly Demand Review Schedule

	Mon	Tue	Wed	Thu	Fri
Week 1		S&OP			Demand review
Week 2					Demand review
Week 3					Demand review
Week 4					Pre-S&OP review

Sample month where the first day of the month happens to be a Monday.

to keep people out of the meeting for any reason other than to minimize their time commitment. Anyone who wishes to participate and has time is normally welcome.

If someone is unable to make the meeting due to illness, vacation, or customer requirements, a replacement is to be named and attend with full decision-making authority in that person's absence. Once a month, usually the last meeting in the month, the agenda changes in preparation for the soon-to-happen S&OP process review. The meeting, just prior to the S&OP, is the monthly demand review where the monthly results will be reviewed to assess total deviation from the original monthly plan as it was locked for review at the beginning of the month (see Table 4.2) This meeting is also sometimes called the pre-S&OP process. The late Oliver Wight (consultant and material requirements planning guru) was probably the first to give it that simple and logical label. Regardless of what you call this meeting, the weekly demand review prior to the S&OP needs to have the monthly emphasis not necessarily part of the normal weekly meeting. The master scheduler gets a chance at this meeting to prepare the organization for better questions and answers at the S&OP meeting coming up soon after. The emphasis is on shared goals.

The agenda for the pre-S&OP meeting is not so unlike the weekly demand review. The biggest difference is twofold: (1) the discussion around demand and production plan variation is more of a summary for the month rather than just the last week's variation and (2) there is a full acknowledgment that the S&OP meeting is only a few days away and questions need to have both answers and proposals for improvement. This acknowledgment makes the meeting more productive by knowing that there is a deadline pending. With everybody in the room to make important decisions, the process can be quite efficient.

Process Ownership in the Demand Plan

The process ownership in the demand planning space can differ from business to business, again because of differences in organization structure and job titles. In the business I grew up in, the vice president of marketing, Margaret Gallagher, had all the product managers reporting to her. Product managers had a huge influence on the business and were aligned by product family — the same divisors that are used in the S&OP process and demand planning. Product managers determined marketing plans and helped manage the sales force policy at the distributors, which were independent dealerships. The product managers were responsible for gathering information from the dealers, massaging it to their liking, and delivering forecasts to the vice president of marketing. In turn, she would deliver these estimates to the S&OP process, where the records would be reviewed and blessed accordingly. She was ultimately the process owner for demand planning, although she got lots of help from the vice president of sales, and both were obviously present for the demand accuracy review during both the end-of-month demand review and the S&OP.

In some smaller businesses, the vice president of sales is also responsible for marketing. In these businesses, everybody from the demand side of the business reports to the vice president of sales. There is little question who the process owner is in these organizations — the vice president of sales. One approach to avoid, if possible, is to delegate process ownership for demand planning accuracy too far down the organization. It is the president or CEO's job to ask the tough questions. The vice-president level is normally the appropriate level for process ownership to reside for the demand plan accuracy, and this person answers to that accountability. It takes a lot of help from the organization to get it right, and the more handshakes there are, the more likely there will be a successful process.

Process ownership means that the "buck stops here." Accuracy is the goal, and the process owner is responsible for providing evidence of both learning and actions to improve the existing demand accuracy by product family. This is a role played out at several management system events throughout the month, the weekly demand review, end-of-month demand review, and the S&OP meetings.

The Measurement Process for the Demand Plan

The measurement process for demand planning accuracy is simple. *The calculation is always from the product family level data and normally has an acceptability threshold of 90%.* The measurement is the average accuracy by product family. Some people become uncomfortable with average accuracy because accuracy by its design is depicted as an average. Nonetheless, it is the best way to handle this metric. Tables 4.3 and 4.4 show how demand accuracy is calculated. The data are from the January performance shown in Figure 4.1. The accuracy is determined from the accuracy of each family. This means the family performance numbers in January from Table 4.1 would calculate per the examples in Tables 4.3 and 4.4.

Table 4.3. Demand Plan Accuracy Calculation

	Accuracy
Product family 1	100
Product family 2	97
Product family 3	90
Product family 4	92
Product family 5	100

Total performance = (100 + 97 + 90 + 92 + 100)/5 families or 95.8%

One of the early questions about Class A ERP measurements in the demand planning space is interpretation of the measurement rules. The forecast and the actual are not always what they seem. For example, most companies have blanket order agreements with their best customers. This just makes good sense. It eliminates unnecessary documentation and paperwork flow. Many companies make the mistake of using the blanket order as the "actual order" in the demand

Table 4.4. Demand Plan Accuracy Spreadsheet

Date: 2 February 200__

Product Family	All Numbers in Thousands												
	Jan*	Feb	Mar	Apr	May	Jun	Jul	Aug	Sep	Oct	Nov	Dec	Jan
Line one	30	30	20	25	34	35	45	78	67	50	35	35	30
Actual	30												
Performance	**100%**												
Line two	30	27	22	21	20	25	15	20	19	16	15	12	10
Actual	29												
Performance	**97%**												
Line three existing	60	80	86	90	88	89	90	78	65	50	35	30	32
Line three new**								20	25	50	75	80	90
Actual	66												
Performance	**90%**												
Line four	50	51	55	65	64	66	73	78	67	55	56	55	50
Actual	54												
Performance	**92%**												
Line five	5	8	6	7	7	7	8	9	10	9	7	6	6
Actual	5												
Performance	**100%**												
Perf. total	**95.8%**												

* The January column in this example has both planned and actual numbers from the month's performance; the balance of the months are planned quantities.

** New products are generally reviewed separately from existing products due to the high-risk opportunity.

planning metric. This often does not make sense because customers normally give a blanket order and then change the quantities or schedule just prior to the ship date. Most of the time, blanket orders are just forecasts. No argument from the point that these signals are pretty firm; in fact, most supply chain agreements include obligations for product on blanket up to certain limits. Nonetheless, these signals really are not much more than a "heads-up" that the customer is about to schedule something. It is good to be ready with some anticipated inventory or capacity.

The metric is designed to measure accuracy of the forecast. Unless you have a special understanding with your customers that makes their blanket orders firm, acknowledge blankets only as forecasts for the measurement. The real order is the firmed up schedule a week or so prior to shipment. At one automotive supplier in Europe, Volkswagen would issue a blanket purchase order several weeks out. Each week, both the blanket and the releases from the blanket were updated. The sales organization in the supplier company was responsible for forecasting how many firm released units per product family the manufacturer would get scheduled by Volkswagen for the month, regardless of the blanket order quantities received. The closer you get to the end item consumer in the food chain, the more applicable this rule will be. The reason is simple — unruly customers.

Demand planning is a monthly metric. The demand plan is updated as required throughout the month, obviously, but for measurement purposes only, the "plan of record" remains locked. This forecast is locked at the S&OP each month and is measured for monthly accuracy by product family. Reporting is only normally done in percentage format at the end of the month demand review and the S&OP meeting. The focus should be on not only accuracy but also what can be learned form the inaccuracies experienced. *Demand planning in a high-performance or "Class A" ERP organization is about shared goals, process ownership clearly defined, and a management system called the S&OP and the weekly demand review to keep everyone communicating properly.* The top management planning processes will be discussed further in Chapter 8.

There is a lot of emphasis on the demand plan from the perspective of the master scheduler. This means it is especially important for the master scheduler to be closely aligned with the demand side of the business. In some organizations, the demand manager and the master scheduler are physically located in the same area. This can be helpful.

It is now time to move into the master scheduling space to look at the details behind good process. In the next chapter, we will focus entirely on that topic.

5

THE DETAILS BEHIND MASTER SCHEDULING

Master scheduling is the process of turning the top management plans into details that can be manufactured in the proper sequence to maximize both efficiency and customer service. This sounds like a tall order and it is. A major software supplier advertises that its software eliminates the need for a master schedule. I find this very interesting, especially since I have some familiarity with the software. While interesting, this concept is not realistic. In businesses that are changing, there is both the need to massage the master production schedule (MPS) and a requirement to align mix according to changes in the market expectations. While much of this can be done through spreadsheet capability, the fact remains that it still needs to be done. Technology may someday replace much of the master scheduler's functional duties, and there is no argument that it is gaining on this objective every year, but at the end of the day, the MPS will still remain. Even in a kanban fully flexible workplace, the MPS still drives available rough-cut capacity and financial plans going forward. *No business runs blind, and the MPS is the lens to make the vision as clear as possible.*

AVAILABLE TO PROMISE

A basic principle of good master scheduling is understanding the simple concept of available to promise (ATP). A simple concept, ATP is the linkage between reality and customer promises. Years ago, ATP was especially important. In

Table 5.1. ATP in an MTO Environment

| MPS Part No. 123-456 | Period | | | | | | | | |
	1	2	3	4	5	6	7	8	9
Item forecast	10	9	9	7	3	3	9	10	12
Actual demand	2								
Master schedule	10	10	9	9	0	0	10	10	10
ATP	8	10	9	9	0	0	10	10	10

those days of long lead time and the strict use of material requirements planning to drive "push" signals of demand forward, ATP was invented to minimize poor promises to customers. Long lead time made it difficult to meet all customer requests when they deviated from the forecasted demand requirement. Customer service in those days was measured and focused mostly on customer promises, not customer requests.

ATP is an acknowledgment of what resources and materials are not spoken for and therefore available in any particular time frame. The calculation is illustrated in Table 5.1 in a make-to-order (MTO) environment. As you can see from the calculations, the ATP assumes consumption of the previous order scheduled, so in period 3, since there are nine units scheduled and no actual demand in that period, the assumption is that all nine are available and none from prior periods are available.

Table 5.2 looks at another possibility. If a customer called in a special requirement out in the future and firmed four units in period 4, the ATP would go from nine to five because four of the nine units scheduled in period 4 are now spoken for. This is confusing to some as there are several unspoken for units available in prior periods. The MPS should only schedule units that have anticipated demand or some build-ahead strategy based on forecasted demand. Given this, the MPS should have inventory available at the end of each period only because of build-ahead plans. Normal demand should otherwise consume

Table 5.2. ATP in an MTO Environment

| MPS Part No. 123-456 | Period | | | | | | | | |
	1	2	3	4	5	6	7	8	9
Item forecast	10	9	9	7	3	3	9	10	12
Actual demand	2			4					
Master schedule	10	10	9	9	0	0	10	10	10
ATP	8	10	9	5	0	0	10	10	10

Table 5.3. ATP in an MTS Environment

MPS Part No. 456-789	Period								
	1	2	3	4	5	6	7	8	9
Item forecast	6	10	10	20	20	20	10	10	10
Projected on hand*	10	10	20	20	20	10	10	10	10
Actual demand	4								
Master schedule	10	10	20	20	20	10	10	10	10
ATP	16	10	20	20	20	10	10	10	10

* On-hand starting position is 10 pieces.

all available MPS items scheduled. This logic makes it unwise to show accumulative availability in the calculation.

After examining one more example, this time in a make-to-stock (MTS) environment, we will move on to more important master scheduling concepts. Notice in the example in Table 5.3 the ATP and the projected on hand are equal. This is because the MPS in this example is scheduling build during each period. If the MPS was only scheduled every other period, as it was in earlier business models prior to lean ways of thinking, the example might look different. In Table 5.4, the MPS quantities are more in the traditional mode of periodic larger quantities. Most organizations are leaning back their MPS quantities and making more frequent orders. Note that in this model, the forecast is decreased by the actual order quantity received. This happens in the software calculation so that the demand quantities are maintained properly. Period 2 demand was 10 and has been decreased by the 3 actual requirements received, now making it 7.

For ATP to be effective in any organization, there are some factors that come into play. The order management team needs to understand and acknowledge the process of ATP. It is easy to always say yes to the customer, but if

Table 5.4. ATP in an MTS Environment

MPS Part No. 456-789	Period								
	1	2	3	4	5	6	7	8	9
Item forecast	6	7	10	20	20	20	10	10	10
Projected on hand*	0	20	10	30	10	10	0	20	10
Actual demand	4	3							
Master schedule		30		40		20		30	
ATP		27		40		20		30	

* On-hand starting position is 10 pieces.

the process agreement at the company is not in alignment with these promises, everybody loses — order management, master scheduling, and the customer! In some organizations, the order management team does not acknowledge ATP and instead prefers to make promises based on "how things have been going" or, even worse, "standard lead time." Although many will advertise it (and even successfully execute it), there is really no such theory as "standard lead time," at least not as an automatic delivery. Standard lead time assumes that orders are received at the same rate at which orders are produced — all the time. This is unlikely. An example of standard lead time at a manufacturing company is when all products in one product family are promised for a three-day delivery. This could be true most of the time, especially in an MTS environment where safety stock is adequate for most demand spikes, but let's talk about a specific example that happens from time to time.

In the situation in Table 5.5, it is easy to see that there is plenty of stock to meet the necessary demand requirements for customers. The safety stock is planned at five days worth of demand or 500 pieces. If the spike in demand is within the safety stock levels, life is good.

In the example in Table 5.6, the same situation exists except that the customer comes back late in the day with a huge unplanned order for an additional 400 pieces above the previous 150-piece order. In this example, the demand exceeds the available stock. In some organizations, the order management group would ignore this fact and let the operation scramble to make the delivery schedule. A possible scenario is that the customer got a surprise opportunity with the stipulation that the order had to be delivered quickly. Normal drumbeat requirements of 100 a day were spiked for this incremental demand.

If this were the case, the best situation would be to evaluate the ATP and for order management to discuss possible solutions with the master scheduler. *If it is impossible to react in the same day (this scenario was stated as happening at the end of the day), the truth should be communicated to the customer.*

Table 5.5. Example Situation

Company X sells product A as an MTS (stocked) product. Product delivery is promised within 24 hours (same day). Demand and plan are adequate.							
	Day 1	Day 2	Day 3	Day 4	Day 5	Day 6	Day 7
Stock (family 123)	500						
Planned replenishment		100	100	100	100	100	100
Forecasted demand	100	100	100	100	100	100	100
Demand from customer Y	150						

Table 5.6.   Example Situation

Company X sells product A as an MTS (stocked) product.
Product delivery is promised for same day.
Demand and plan are not adequate.

	Day 1	Day 2	Day 3	Day 4	Day 5	Day 6	Day 7
Stock (family 123)	500						
Planned replenishment		100	100	100	100	100	100
Forecasted demand	100	100	100	100	100	100	100
Demand from customer Y	550						
Shortfall	50						

One possible scenario is as follows:

1. First the customer uses some of its 100-a-day "normal" demand inventory to cover the shortfall of 50 pieces.
2. Then the operations team at the supplier immediately goes into overtime mode to backfill the safety stock and the next day's normal requirements.
3. Lastly, the 50 extra pieces are promised for day 2, with all other requirements met same day.

The risks are great with this scenario, and if there are other customers that use this same part, it is not *necessarily* the best solution. That is why it is so important to make the promise decisions with acknowledgment of stock. Sometimes rules can help make the exception decisions easier.

RULES OF ENGAGEMENT

Rules of engagement will be discussed several times as the topic of master scheduling continues to unfold. These rules govern the decisions to service customers as well as manage costs and risks. For example, using the scenario from Table 5.6, there could be a rule governing promises that says:

> For MTS product family 123, promises will be made from inventory up to order quantities of 50 pieces. Single orders received for more than 50 pieces will be considered "special orders" and not planned for or inventory stocked in readiness for normal demand. Master scheduling will work these exceptions and respond to order management with a promise within two hours from receipt of request. This policy will be reviewed monthly and is the responsibility of sales to revise.

In reality, these orders can be shipped in many cases. The solution comes from, again, agreement on cost and customer service. Some customers are always worth the extra cost; others may not be. By understanding the "math behind the process," better decisions can be made to minimize inventory exposure and maintain high service levels. *It is important to note that rules of engagement are not made to say no to the customer. These rules are designed to understand and acknowledge the costs of saying yes.*

The master scheduler is the keeper of the rules. The demand side of the business has its eyes and ears closest to the marketplace and must make sure that the rules are in best alignment with the best interests of the company. *There are few practices that have more impact on the business than properly managing rules of engagement in order entry and processing.*

MAINTAINING THE MASTER SCHEDULE

Decisions on inventory strategy and rules of engagement are the main procedural drivers for master scheduling. This is not where the job stops, however. In fact, there is a lot more to discuss when viewing master scheduling in its entirety. The master schedule is the set of marching orders for the operations side of the business. It is the drumbeat that determines rates and rhythms for each product line. Information is constantly being fed to the schedule. It comes from forecast changes, the Friday demand review, surprise orders, process variation like scrap or rejects, and it can even come from top management priority changes. That makes schedule maintenance a major effort and focus for the best master schedulers.

Maintaining the master schedule means also managing changes to a minimum. In most organizations, change can affect cost, and too much change can create chaos. In one business I worked with that made household products, one popular option consisted of a stainless-steel outer shell. At the time, stainless steel was in high demand throughout the world. China had entered this particular manufacturing market in a big way, and as a result, shortages of stainless steel developed. This put a significant schedule maintenance issue on the master scheduler. Stainless steel allocation in this particular instance meant that knowing how much material would be available would be possible only a few days prior to delivery. Since this plant was a high-volume operation with many other components, constant juggling of the schedule resulted. Costs are a major concern in any commodity market, and this was no exception. The master scheduler had to maintain an open ear directly to the supplier as well to as the market.

In some companies, this situation would lead to friction between sales and operations. After all, to some, selling the product is the difficult task; building it is "cookie-cutter easy." In high-performance organizations, the order management team would have access to product availability, and promises would not be made in quantities or time frames above the ATP. This requires good teamwork and trust that the master scheduler will build all required products that are possible and drive material availability with a high priority.

The master schedule in most businesses is maintained globally on a weekly level (at a minimum) and massaged throughout every day. This maintenance requires some rules to be addressed and understood.

Past Due Requirements

In reality, there are no such thing as past due requirements in a well-managed MPS process. This is not to say that things always go well and nothing ever slides past the original due date. Good master scheduling process requires that all dates be as accurate as possible. *When dates get out of sync, they need to be realigned. Good master schedulers realign the MPS at least once a week to synchronize the schedule with reality.* If realistic dates are the objective, past due requirements will always be maintained out of the schedule. After all, past due means something is *going to happen yesterday*! That is not likely, at least not using normal logic. In some companies, management gets very nervous when past due schedules are rescheduled. It is like giving up or giving in to temptation. Past due signals drive inventory that will not be used in the time frames given, which makes the component and raw material piles bigger and harder to sort to get the items really required in the receiving department and in assembly. *There is no logical reason to keep past due dates in the MPS — realign the MPS regularly!*

Building the Plan

The MPS is made up of both forecasted or planned orders and actual orders. In an MTO environment, the cadence for the drumbeat in a high-performance organization is extended out to a 12-month rolling plan (this would also be true in MTS or assemble to order [ATO], but we will use MTO for this specific example). Most of this would be forecasted demand, unless you are building airplanes, submarines, or some other long-lead-time item. Because of this, it is necessary to realign the MPS as forecast is consumed by actual orders. In the short term, the time buckets in this 12-month rolling plan would be in weeks and in the closest time frames would most likely be in days. In Figure 5.1, you

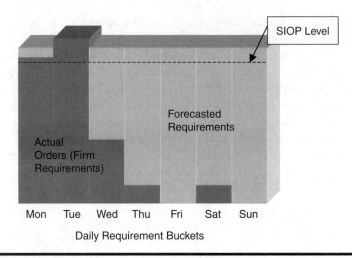

Figure 5.1.   Master Schedule Actual and Forecasted Requirements

can see that the actual orders do not necessarily all come from current period requirements. Sometimes customers want to have product shipped in the future and give notice accordingly. Each situation has to be covered individually and specifically.

The plan should correspond exactly with the sales and operations planning (S&OP) plan. This means that each month, out 12 months rolling, the S&OP plan would show the same number of units as the MPS (see Figure 5.1). The S&OP process is normally built in product families, not exact stockkeeping units (SKUs). In many of the best-performing companies, the forecasted orders are also loaded in product families, not specific SKU designations. Normally, these requirements are driven from planning bills. When this happens, the master scheduler, depending on the enterprise resource planning (ERP) software capability, must maintain the schedule by translating the actual orders into forecast consumption.

To better explain this, think of an order as part 123-456. The planning bill of material (BOM) was set up to drive the family of parts with the components of part 123-456 listed (see Figure 5.2). The top-level planning BOM is not the same item number (123-XXX) as the actual order (123-456). In this case, if the software is not set up to consume forecast by the family group, both the firm and forecasted orders could continue to drive requirements in the ERP system, resulting in too much inventory and wasted capacity.

In these cases, the master scheduler may massage orders by keeping track of current orders and decreasing the proper product family by the appropriate

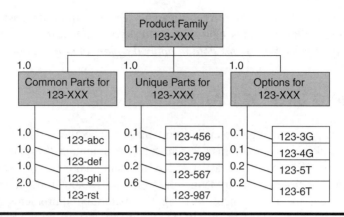

Figure 5.2.   Planning BOM Example

sum. This sounds like an insurmountable task, but in most environments it is not difficult; it just requires disciplines and predictability in the tasks each week. As you can imagine, if the schedule is shuffled frequently, more confusion and workload issues can result. The best high-performance organizations carefully weigh the costs and benefits of changes to the build plan but never allow the MPS to be out of line with the actual build.

In the example in Figure 5.2, the planning bill would be driving product family 123-XXX. The actual order might come in with a different part number, such as 123-456. Some ERP systems do not recognize the relationship to allow for the consumption of forecast to happen automatically. This makes it imperative that the master scheduler maintain the schedule frequently. Depending on volume, this might be every hour, every day, or in lower volume instances every week.

The plan built from the S&OP results would usually start with a product family plan. The master scheduler would typically build planning BOMs based on three data elements:

1. **Historical data on mix** — This information shows how the normal mix might happen (such as 67% of sales of this car model normally have V-6 engines).
2. **Information from marketing and sales on mix going forward** — This might be expectations of changing mix due to promotions, the introduction of new products, etc.
3. **Seasonality data on mix** — Most companies have some seasonality. Seasons or holidays often are a factor in demand and mix.

When this information is current, it can be very helpful in the master scheduling process. Planning bills can be built to drive planned requirements. *The S&OP is the source for volume data to apply the planning bill mix percentages.*

PLANNING BILLS OF MATERIAL

Planning bills are a huge asset to good master schedulers. These planning shortcuts make it easier for master schedulers to maintain the plans without thousands of SKU-level forecasted orders to manipulate. If, for example, in January the forecasted plan for the month of July has 100,000 units and planning bills are used, the requirements driving the material requirements planning (MRP) plan would consist of one BOM (the planning bill) with a quantity of 100,000. The planning bill would take the expected mix and ratio it (see Figure 5.3).

In this example, the common parts are separated from the rest of the possible changing components for product family 123-XXX. This ensures that if some buffer is required for mix purposes, the common parts will not also be buffered at the same level. The S&OP plan already takes this into consideration, and there is no need to double up on buffer stocks. If the top management S&OP plan calls for 100,000 units to be planned, that is all that should be planned — no more. Mix components, however, create a different issue. If the S&OP plan calls for 100,000 units in the month of July, there is no further definition of the detail requirements. There could be several different SKUs offered within

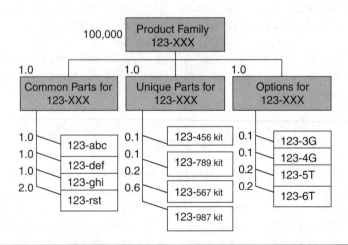

Figure 5.3.   Planning BOM

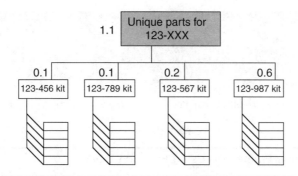

Figure 5.4.   Unique Components within a Planning BOM

that one product family (123-XXX). This is where the planning bill comes to the rescue.

Let's say, to keep it simple, that there are four different configurations (SKUs) within product family 123-XXX. The SKU-level items are 123-456, 123-789, 123-567, and 123-987. In Figure 5.4, these are listed under unique parts for 123-XXX. In that section of the BOM, each configuration's unique parts would be loaded as a kit without any common components.

If the SKU-level mix within the product family can migrate up or down, as is the case with most product families, buffer may be called for, especially if there are long lead times for components or short lead times for customers. In the case of 123-456 in Figure 5.4, the mix is normally 10% of the total requirements for the family. The master scheduler loads the requirement ratio at 0.1 or 10%. This action does not create buffer in itself. Buffer can be controlled easily at the top level (123-XXX unique components BOM). At the top of Figure 5.4, there is a designator of 1.1 or 110% for this BOM. This will automatically drive components at a 10% buffer and will fluctuate with the S&OP plan as it is maintained at the top level for each month. By driving only the unique parts at a "required plus buffer" level and common components only at required levels with little or no buffer above the S&OP agreement, the flexibility is maintained with the least amount of inventory. Maybe most importantly, management is now in control of the inventory rather than each planner.

In some organizations, the opposite is done. Common parts are seen as low risk and are driven at buffer levels, and unique parts are planned at minimums. Flexibility and cost both become problems in this scenario as mix changes and expediting becomes the norm. Inventory is also not as low as it could be. *Master schedulers need to remember that the decisions for buffer are really made at the top management S&OP process in most high-performance organizations.*

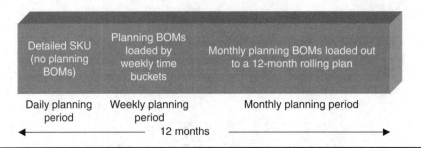

Figure 5.5. Time Frames for Using Planning BOMs

The master scheduler is the one who develops the plan for the top management S&OP anyway and should have total confidence in the plan.

Planning BOMs are a helpful tool for master scheduling and should be in the master scheduler's toolbox. They are used in different manners in different time frames (see Figure 5.5).

In the daily planning period, the requirements are normally at the SKU level based on actual orders loaded, either planned or firm. Depending on the business, this can be a few days or several. It depends on inventory strategy, lead time for components, and lead time requirements from the customers. The philosophy used is the same, but with a little different configuration in each company. By using the planning BOMs in the future time periods, it becomes a much less maintenance-intensive process.

Obviously, the mix ratios in the planning BOMs make a difference in how much inventory is being driven. It is also reasonable to ask what to do about the fact that as products move through their product life cycles, the mix can change. This is absolutely true and also needs to be part of the thought process as the master scheduler communicates with marketing and sales. Some master schedulers use a certain time each month, maybe the third demand review of the month (Friday meeting), to review ratios with the demand manager. Remember that the good news with MRP is that it will not duplicate orders. If the planning BOM drove requirements that have not been consumed, MRP will net against the on hand and will not require more to be ordered. This is not a ticket to leave the planning BOMs unattended. Quite the contrary, but it is a part of the logic for using the time-saving planning BOMs.

The S&OP gets updated every month, and often there are changes that need to be made out in future months to keep the master schedule in synchronization with the top management S&OP plan. *By using the planning BOMs, the changes can be made without getting into the details — provided the S&OP, master schedule, and planning BOMs are all in the same product family groupings.*

Therein lies a good lesson. It is imperative that these three processes share the same product groupings. It is also just as important to include the operations plan and business plan in this scenario. These top management plans should also share the exact same product family configurations.

TWO-LEVEL MASTER SCHEDULING

The process of master scheduling inventory strategies other than MTS product is sometimes referred to as two-level master scheduling. This term comes from the fact that the MPS items configured in the planning system are not the customer-ready finished goods SKUs in ATO or MTO environments. Instead, the MPS items in these manufacturing environments are planned at least one level down from the level "zero" finished goods level. The sublevel items (often subassemblies or fabricated SKUs) are scheduled by the MPS to be ready in anticipation of receipt of the customer purchase order. Once the purchase order is received, components for the specific customer-required configuration are assembled and shipped to the customer. The final schedule to make this happen is sometimes referred to as the top-level MPS or sometimes as the final assembly schedule. It is the two levels, MPS and final assembly schedule, that make up what is referred to as two-level master scheduling.

MATERIAL REQUIREMENTS PLANNING

At the heart of almost every material planning process within manufacturing is some form of MRP. MRP has gotten a bad rap in recent years. *Among the "constraint" advocates, Six Sigma people, lean gurus, software visionaries, and the just plain inexperienced types, the word on the street is that MRP is dead. Nothing could be farther from the truth.* When the discussion point is strictly described as "netting of future requirements," few would argue with the need to do this in manufacturing planning both now and in the future. The truth is, almost every major successful manufacturing business on the face of this planet uses some form of MRP, whether they want to admit it or not.

The software vendors have as much to do with this confusion concerning the terminology as anybody, and it is understandable. Much like the label "station wagon" is out of date (the term is now SUV), MRP also has a few new names. Most of these names are recognizable by hints like "advanced scheduler," "advanced planning and scheduling module," or even "drop-and-drag order sequencer." These newer labels are the "four-wheel-drive SUV" versions of the MRP station wagon! Some of the names are actually quite clever. At the

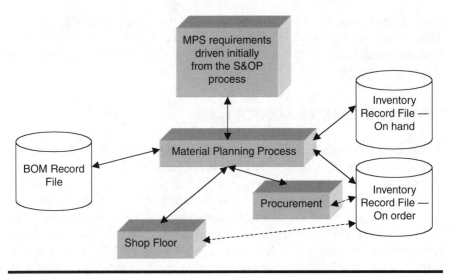

Figure 5.6. MRP Inputs and Outputs

end of the day, however, these planning engines are basically netting requirements and time phasing the resulting signals for either procurement or manufacturing — MRP!

Material planning is the process of taking the requirements handed off from
the MPS and determining what if any components need to be ordered versus
those already available. Availability is derived from data associated with on-
hand stock, on-order quantities from procurement, and/or on-order/in-process
manufactured item quantities (see Figure 5.6).

The MPS requirements are the real drivers of material planning. The master
schedule is massaged continuously to synchronize the supply side with customer demand. The material planning engine in the ERP software business
system is the calculator that sorts out what to do with these MPS signals. For
the rest of this discussion, MRP will be the label referred to when describing
this detail-planning engine, even though some software will use a different
label.

DATA ACCURACY

MRP has several activities to carry out once the MPS requirements are refreshed. The first activity is to pick up requirements from the BOM record for
each requirement. The BOM record holds several pieces of important information, including:

1. Top-level part number (parent-level number)
2. Component part numbers called out in the upper level
3. Usage per component in the upper level assembly
4. Unit of measure for each component
5. Lead time and lead time offsets for each component

When the BOM information has been accessed, the next information the MRP engine requires to complete its job is the inventory record. There are two types of inventory records for this discussion: on-hand inventory and on-order inventory. The names are descriptive enough, but for clarity there is an important Class A ERP influence in this space. Data accuracy becomes very important as realization of data dependency is underlined. For this planning engine to do its job effectively, the BOM records must be impeccable and inventory records must be pristine. Class A ERP, or high-performance criteria in general, requires BOM accuracy to be at a 98% minimum and 99+% to be considered good. Inventory record accuracy (accuracy of piece-part balance by location) needs to be 95% to meet a threshold of acceptability and 98% to be considered robust. These are important requirements to keep process variation and associated costs to a minimum whether you are looking to become Class A ERP certified or are just interested in high-performance ERP.

If the master schedule is no better than a 90% level of accuracy, which means that the requirement fields are about 90% accurate in either quantity or date, MRP will be negatively affected by the 10% variability. If the MPS process variation exposure is increased by having less than 98% BOM accuracy or if the inventory record accuracy is below 95% (piece-part balance by location), the accumulation of process input variation becomes an exponential problem affecting the output accuracy of MRP. This is not rocket science. It is simple logic. The goal of any organization should be to use the business system on autopilot at every possible opportunity to lower cost. Bad data eliminates using autopilot because human horsepower is immediately and consistently required to offset the process variation caused by inaccuracies in data. *Too many organizations today still do not appreciate the huge payback available from such simple efforts as process transaction disciplines.*

ORDER POLICY DECISIONS

MRP results in signals the planners can respond to, according to the calculation resulting from inputs and outputs. When a requirement is determined for a particular component, one of the first questions the planner has to answer is how many of this component to order. Sometimes the decision is easy (for example, only the number to match requirements), but other times it is not as simple. Consider, for example, a situation where an inexpensive item is purchased from a supplier located thousands of miles away. If the requirement is for stainless steel bolts at a quantity of 17 per assembly with a price of 30 cents each and the source for these bolts is China, what is the right quantity to purchase? The answer, even given the data of price and source, is still "it depends." It depends on inventory strategy of the item and the upper level requirements, and it depends on anticipated usage beyond any known history. Order policy begins to address the issues and answer the question "How many should we order?" There are a few types of order policies that are used often and one (economic order quantity) that is not but occasionally gets press time:

1. **One for one** — This is the just-in-time, lean, kanban approach. No extra are ordered, and nothing is ordered until there is a customer demand that creates a requirement. If 17 are needed, only 17 will be ordered. This approach is used in an engineer-to-order (ETO) or MTO inventory strategy. In a lean environment, most finished goods are made on either an ETO or MTO strategy.
2. **Lot for lot or order for order** — This policy is a deviation on one for one. In this situation, subassemblies may be requisitioned in anticipation of a customer order and require components. Items with this order policy assigned are not ordered unless there is a requirement from another parent level. In that case, the quantities are requisitioned to match the upper level lot size.
3. **Time value or fixed period order policy** — With this policy, parts are ordered in quantities to cover the usage for a certain period of time, often a week or more. If the assigned time value for a part is two weeks' worth and the usage is expected at three per day, for two five-day weeks, the order size would be 30 pieces (two weeks times five days times three per day).
4. **Fixed order quantity** — This policy dictates certain predetermined quantities. In some process flow businesses, these quantities are also referred to as campaigns. There can be many reasons for this practice, such as how many can be made from a sheet, how big the vat in the process is, what the normal demand for this item is, how many fit on a pallet, etc.
5. **Economic order quantity (EOQ)** — Very few organizations use EOQ today. The formula is the square root of 2AS over IC, where A = cost of an order to process, S = units per year, I = inventory carrying charge, and C = unit cost (see Figure 5.7). Most professional materials experts believe that while the *theory* of EOQ might be valid, the application of

Figure 5.7.   EOQ Formula

it is not. The main problem with it was the ability to adjust all the variables and end up with whatever lot size you wanted.

ABC STRATIFICATION

Along with order policy code, stratification also plays an important role in ordering inventory from MRP signals. Inventory stratification is the segmentation of items into layers by monetary or usage values. Vilfredo Pareto's 80/ 20 rule works well here. Generally, the dividing layers are done by:

> A = 75% of the monetary value of inventory
>   10% of the item numbers
> B = 15% of the monetary value of inventory
>   10% of the item numbers
> C = 10% of the monetary value of inventory
>   80% of the part numbers

Some organizations also have a D category that divides the C items into two groups. D items are usually components that are *very* inexpensive (such as common small washers, screws, etc.), with values much less than one cent each.

Most ERP business systems can automatically calculate the inventory layers and give the user parameters that can be set as desired. Stratification is done to make sure there is more emphasis on the items that have the most risk associated with them. Therefore, critical components that are difficult to get or are allocated are forced into the A item code regardless of the ABC calculation.

RUNNING THE MRP NET-CHANGE CALCULATION

Since MRP was invented about 30 years ago, a lot of progress has been made in both software efficiency and hardware capacity. The rule of thumb in the beginning of MRP application was to run the regenerative net-change program and calculation every weekend. That was because the software program and corresponding calculations took hours to run, and often nothing else could be processed while the room-size computer churned the data, layer by layer. Even more interesting was the philosophy that running it too often introduced a large amount of unnecessary system noise (variation) into the manufacturing process. It is actually quite humorous to think about that today. That system noise was driven by changes in customer demand, and the reason it was called noise was because the manufacturing managers did not want to change over machinery setups in the shop. They wanted to run long lot orders. If MRP was not run very often, changes still happened but they were not communicated to the shop, which artificially created stability in the schedule. This luxury was offset by huge amounts of inventory in finished goods to avoid this "noise." Expediting was often funded willingly, and heroes were created in the shop and in purchasing daily as companies like FedEx became household words. Today, things are very different, from the scheduling philosophy standpoint.

Customer demand is why organizations are in business. Ignoring it is foolish and costly. If the existing production process is not matched to the present customer behavior and demand, it is time to reevaluate the manufacturing strategy. That would include inventory strategy and market offerings. When these strategies are in line with market need, how often you run the MRP program is a nonissue. The more it is run, the more accurate the current schedule is, and life is good.

Most high-performance organizations run MRP at least daily, and many run it several times a day. Again, if the inventory strategies and rules governing the supply chain are adequate for the markets you have chosen to serve, there will be nothing but gains from more frequent MRP runs.

TIME FENCES (SEE FIGURE 5.8)

Fixed Fence

When suppliers make only enough to cover the fixed period in the short term, this can mean overtime or special freight when changes are made to the schedule within this short-term horizon (this fixed fence is often hours or maybe a couple days at most). In the best examples, the customer picks up charges within

Figure 5.8. Typical Time Fence Agreements

this fence. After all, the idea within this time fence is to keep the process variation and associated costs to a minimum. Unneeded inventory is unneeded cost. A common scenario has the supplier focused on exactly what the fixed schedule calls for and stocking only a couple of units (or components for units) over the demand. This gives some flexibility, but for cost reasons is limited within this short period. A company in Mexico that supplied components to Caterpillar had agreements with its customer that the fixed fence would be two days. Even within that short schedule, quantities would change at times. This supplier had a few extra assemblies for all of the most popular items available all the time. A pull system kept this buffer stock filled. Caterpillar knew the buffer quantities, and a handshake rule allowed changes up to the buffer level. This type of agreement works quite efficiently.

Firm Fence

This next fence requires a handshake with the supply chain and might allow up to ±20% on requirements communicated at the beginning of the period. For example, in an environment with a 3-day fixed fence and an additional 7 days in the firm fence, on January 1 the schedule says 100 units per day of a specific part number through the entire 10 days. The flexibility agreement within the fixed fence may be plus or minus two units per day. On the fourth day out, flexibility requirements are increased to (possibly) ±20 units per day. These kinds of handshakes allow everybody to develop the right inventory strategy and keep costs to a minimum. These rules are like risk management for the supply chain. It is in the interest of both parties to have the agreements at a sensible level. The wider the schedule swings can be, the more capacity needs to be committed, which drives costs. This is also why it is so important that the schedules be as accurate as possible.

Weekly Rate Fence

It gets easier to define flexibility the farther out you go on the timeline from delivery. Beyond the initial barrier of the firm fence, the master scheduler would normally schedule detailed daily requirements out for some reasonable time frame. If the fixed period was 3 days and the firm fence another 7, the weekly schedule fence would start after that, maybe from day 11 to day 60. This means that from day 1 to day 60, all the requirements in the MPS are scheduled in weekly buckets with very specific SKU information driving the supply chain. The flexibility rules keep getting more demanding the farther from the current period you go. Many times, the rules of engagement with suppliers can increase flexibility requirements to ±20% out beyond the firm fence.

Beyond the daily schedule in the MPS, master schedulers will normally transition to weekly rates. Usually the farther you get from the current period, the less accurate the requirements are anyway. Keeping the level of detail within the schedule at the SKU makes a lot of unnecessary work that will probably be changed several times anyway. It is often much more efficient and accurate to use planning bills and insert weekly rates by product family for the next few weeks. Flexibility agreements are often wide open at this point and are a matter of negotiation since the period is often outside accumulative lead time for the supplier anyway. It is normally in the best interest of all parties to increase as required, especially when there is time to prepare for the increase.

Monthly Rates Beyond Weekly Rates

For the same reasons master schedulers move to weekly rates at some point within the 12-month rolling horizon, out at some point the level of detail can be minimized further. This is done by simply moving from weekly rates by product family to monthly rates by product family. This keeps the maintenance work to a minimum and allows for easy alignment with the S&OP process monthly. Again, flexibility is in open season. At this point, the time fence continuum might be better visualized as a wider and wider flexibility requirement.

In Figure 5.9, the fences increase in flexibility each time a fence line has been crossed. At each fence, the flexibility requirements increase by the amount that the handshake allowed. This allows each party to minimize its costs by keeping the inventory on one side of the agreement. In the fixed fence, very little extra inventory is held by the supplier, whereas the supplier increasingly takes on the flexibility requirements as the timeline increases. This does not necessarily mean inventory, but it does mean that planning has to take place

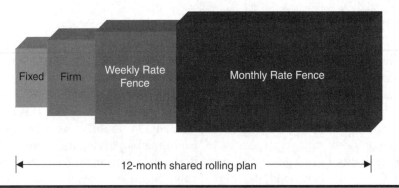

Figure 5.9. Expanding Fence Flexibility Requirements

at the supplier to facilitate possible scenarios. *When these handshakes are in place, both the supplier and the customer understand the risks and can plan for them accordingly.*

THE MASTER SCHEDULING HIERARCHY: CENTRALIZED VERSUS REMOTE

One of the questions almost always asked at one time or another in a manufacturing company with multiple plants is whether the master scheduling should be done remotely in the plants or at central command at corporate. The answer is normally very easy. There are very few examples of centralized scheduling done well; having said that, it is possible to do it (see the interview with Robert Turcea in Chapter 17). The truth is that in applications where centralized influence is done, there is often, but not always, a person in charge of scheduling "tweaks" on-site. Let's look at the two possible scenarios.

Centralized Master Scheduling in a Multiplant Environment

Centralized master scheduling can take various forms. This book will not try to cover every possibility but instead will cover the basic premises for success in this mode. The more complex the product and the more local the plants are to suppliers and/or customers, generally the more scheduling horsepower is required at the plant level. In situations where the products are common between plants and are only manufactured in specific geographic areas because of transportation costs for finished goods, the master scheduling configuration is often a centrally controlled process. Examples of this type of product might include baked food products (such as doughnuts, potato chips, or ice cream cones), corrugated packaging materials, or plastic containers where common molds occasionally are shared between plants. There are other situations where centralized scheduling makes sense, but this rule covers the lion's share.

Centralized planning and scheduling in its fullest form means that the central master scheduler is actually developing the sequence and schedule for each plant and product expectation. Finished goods are usually planned at the geographic level, which means that demand must be also sorted by geographic requirements. Centralized scheduling usually requires centralized sales management as well as centralized order management. Figure 5.10 depicts a typical organization chart.

In this scenario, it is necessary for the master scheduler to keep in close contact with the plants so as to have a high-level awareness of demonstrated capabilities and schedule accordingly. It is typical for the master scheduler to

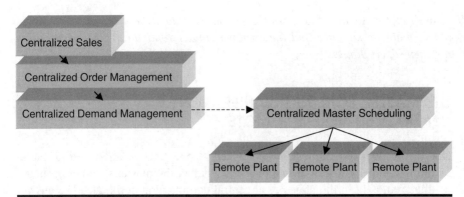

Figure 5.10. Typical Organization Chart with Centralized Master Scheduling

have a weekly clear-to-build process where the plants agree to the schedule ahead of time. This often happens on Friday, with a review completed by the plants in time for the publication of the weekly plan on-line.

Some of the rules that might be used in centralized master scheduling are as follows:

1. The master scheduler must have clear knowledge of the plant capabilities.
2. Preventative or productive maintenance schedules need to be provided to the master scheduler ahead of time; monthly plans work well by allowing the scheduler to fold them into the requirements.
3. Rules of engagement need to be well defined by plant and process. This includes time fences being well defined.
4. The master scheduler must have easy and fast methods of communication to the production management at the plant.
5. The plants must follow the schedules.
6. A shared business system with joint access makes the communication much easier.
7. Fluctuations in plan need to be communicated to the master scheduler immediately. This includes machine-down scenarios, accidents, and other issues affecting plan predictability.
8. Plant capability must be fairly consistent and predictable for centralized scheduling to work well. If deviations are common, it is extremely difficult to control schedules from a remote location.
9. The master scheduler must have control over the schedule, not simply guidance.
10. If services are dispatched from a central gatekeeper, such as all calls for cable repair go into a central location and repair resources are

flexible as required, the master scheduler and the gatekeeper should reside at a central location. A manufacturing example would be transferring tools in a plastics company to other plants when capacity restraints call for them. In that example, centralized master scheduling would be best executed.

Master Scheduling at the Site

The other method of organizing for master scheduling is to have a master scheduler at each site. This is more typical for most complex products such as automotive supply chain or capital goods manufacturing. In this scenario, each plant is a center of demand and supply. Demand management and order management can be centralized, but do not have to be if the products are different at each facility and not closely related. This is the preferred methodology in most instances. Often, there is still some sort of schedule coordinator at the corporate level, but not to the detailed level. This corporate coordination usually is no more than a consolidation of information for the top management S&OP process once a month. Figure 5.11 depicts a typical noncentralized organization chart. Companies also have to consider contact points for customers and, for their ease, keeping these contact points to a minimum

In this scenario, not much has changed from centralized schedules in terms of master scheduling requirements, just the location of the master scheduler.

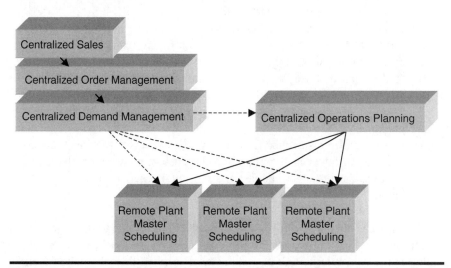

Figure 5.11. Typical Organization Chart with Centralized Sales but without Centralized Master Scheduling

This may make it seem like a good argument for always having a centralized master scheduling process, but just the opposite is true. Generally, the accepted and correct rule is to start with the premise that master scheduling is best done at the plant level. *In most (but not all) high-performance organizations, master scheduling is not centralized, but instead is executed at the facility level.* Although not scientific proof, two of the three companies discussed in Chapter 17 have master schedulers in each facility.

Some of the rules that might be used in remote facility master scheduling are as follows:

1. The master scheduler must have clear knowledge of the plant capabilities.
2. Preventative or productive maintenance needs to be scheduled by the master scheduler and maintenance. Handshakes allow appropriate time for such value-add maintenance activity.
3. Rules of engagement need to be well defined by both demand management and supply-side management. This includes time fences being well defined.
4. The master scheduler is in constant contact with production, often updating metrics for production.
5. The schedule is the only map. Exceptions have to be approved by the master scheduler.
6. Fluctuations in plan need to be communicated to the master scheduler immediately. This includes machine-down scenarios, accidents, and other issues affecting plan predictability.
7. Plant capability must be fairly consistent and predictable for high-performance manufacturing to be successful. In applications where machine uptime is not at least consistent, master scheduling works much better at the facility level. If process deviation or machine capability is in question, it is extremely difficult to control from a remote location.
8. The master scheduler must have control over the schedule, not simply guidance.
9. If services are dispatched from a remote gatekeeper, such as direct communication between the plant and the customer receiving frequent deliveries, the master scheduler and the gatekeeper need to reside at the plant.

As you can see, the rules from one organizational dictate to the other do not deviate much. That is because the job does not change much; only the geography of the scheduler's authority changes. The rules that do change make a substantial difference however. *Designing the organization structure is an important decision in creating a master scheduling success.*

VENDOR-MANAGED INVENTORY AND
THE EFFECTS ON MASTER SCHEDULING

One of the most recent and popular terms in materials and purchasing circles is vendor-managed inventory (VMI). VMI is a logical evolution of consignment inventory. Consigned inventory was a much less rigid but supportive practice where inventory was placed in the customer facility, but ownership was not transferred until consumption took place. Transactions often were only logged after a periodic inventory was taken, usually by the supplier. If inventory decreased, an invoice followed for that amount. Replenishment resulted from the combination of actual consumption and forecast information from various sources. This worked pretty well for the customers unless inventory disappeared without known consumption due to mismanagement of storage within the customer warehouse. One client I worked with had a gold component in its BOM that was under consigned inventory controls. The area was cordoned off with no access except for people authorized to handle the gold product for the fabrication. Most operations are not managed with that kind of discipline.

VMI is a process that taps into the customer's information streams just as if the supplier were part of the customer's organization. There are some real advantages to this approach, not the least of which are the controls on cost. Of course, this is only as good as the data streams, and if the customer has no clue about future demands, not much about the process has changed except the acronym! *The best applications of VMI are with customers that utilize good S&OP processes and have both the demand- and supply-side teams linked into the inputs, metrics, and results.*

In a VMI environment, the product is treated just like any other MPS item. Demand comes from either forecasted or firm requirements and can be driven from planning bills or actual BOMs. The master scheduler often plays the role of the operations manager in the data-mining process by asking the tough questions concerning trends and unexpected demand spikes. Of course, the real goal is to understand if surprises will continue or are just that — anomalies. VMI is normally controlled with a little more personal involvement from the supplier compared to consigned inventory or normal manufactured components. The data source for future requirements is not necessarily as firsthand as demand that comes from the supplier's own S&OP process. The arrangement is also for the benefit of the customer in terms of both customer service and cost. This necessitates good attention to details.

Whereas consigned inventory was normally an MTS inventory strategy, VMI can be other strategies. Using the data from the customer's demand stream, the supplier, and the associated master schedule, it can be built around an ATO or MTO strategy (from the customer's perspective, not the supplier's). If, for example, the customer planned an ATO process with MPS items at the subassembly level, the customer would drive these MPS items to be available or in stock. The same thing would happen in a VMI environment; the subassemblies would be driven from the same information, except by the supplier instead of the parent company. Add shared information and the process becomes somewhat seamless.

In this application, the supplier in essence is just an extension of the customer's organization as it relates to these components. Years ago, Henry Ford used a process that was the ultimate in vertical integration. He processed iron ore into cars, all within his company's own real estate. He may have been way ahead of his time. Evolution of process brought the concept of "core competency" to the forefront, and the result was movement away from vertical integration to suppliers and then partnerships with "experts" in various supply chain areas. *We may be moving full circle by combining our efforts and once again becoming fully integrated, this time with multiple organizations utilized in the process.* The model becomes very much like Henry Ford's, but with divided real estate. There are examples of companies that have their processes and equipment inside the customer's building, getting even closer to the vertical model. Advantages of this new model include sharing capital requirements and sharing risks. There are great benefits in process integration, and with autonomy built into this integration, more accountability and customer awareness are born, at least in the well-managed examples.

VMI is a powerful tool for suppliers and can differentiate one supplier from another. Master schedulers need to be prepared to take advantage of suppliers with this value-add potential and prepared to link with customers to increase value-add downstream. Too many master schedulers view this as complex and difficult to establish. In reality, it is no different once the communication links for data with full disclosure are established. VMI from the supplier perspective and the MPS at that point in the food chain is important as a weapon for competitive advantage.

SUPPLY CHAIN LINKAGE TO DISTRIBUTION

The other direction in the supply chain from suppliers is toward the customer. In many organizations, this is executed through distribution. When distributors are involved, especially independent distributors, the master scheduler is not in control of their inventory strategies and levels. This can be of great concern to the original equipment manufacturer (OEM). The answer is to manager the distributor just as you would a leg of the business to the degree you are able. Metrics are the key.

Independent Distributors

If they want to continue to have access to your line of products, distributors need to maintain a standard of service quality. This requires varying models within varying markets. In one capital goods market, distribution was done with independent distributors through contractual agreements that give them access to the product. The contract documented performance expectations. These expectations included stocking of inventory. This can be especially important when service parts are required. Distributors that do not stock any inventory and therefore degrade customer responsiveness have to be questioned as to the real value they contribute within the supply chain. At one consumer goods supplier, dealers were treated only as customers from the OEM perspective. The result was that many distributors took advantage of the OEM. The customers suffered.

Measurements should be made and progress fed back to distributors on a weekly or monthly basis. Awards should recognize good behavior, and warnings should be given to distributors that do not do carry their load. The master schedule has to consider the risks and customer service in all situations. This might include planning more buffer in areas where distributor quality issues are known, at least until the processes can be repaired.

In-Company Distribution

Another method of distribution is to use company-owned or sometimes company-subcontracted distribution. This normally means that the company has at least some MTS inventory strategy and requires inventory to be positioned near the customer for the best customer service. Consumer goods sold directly to stores can be an example, as can grocery sales or even services provided by organizations such as FedEx. When inventory has to be positioned, forecasts in the demand planning need to consider not only volumes but also geographic demand. It becomes important to have the product in the right place. When it is in the wrong place, it either contributes to increases in transportation costs or the deliver-on-time performance suffers as stock exists but is not at the right location for use. There are a few ways to help manage the distribution. One is demand pull.

Demand pull is the thought behind the Japanese term kanban. It simply means that demand draws from stock in the distribution centers create a pull signal for replenishment. For example, if the distribution center in Los Angeles sells 50 red-and-blue units today, a replenishment signal would go to the manufacturing site for a quantity of 50 red-and-blue units. It is a simple process but one that does not always provide the best service in cyclic or seasonal demand. In these cases, the replenishment quantities would fluctuate depending on the time of year. The master scheduler has to be well aware of these expected demand shifts and plan accordingly.

Another option is distribution requirements planning (DRP). DRP has been around for a long time. It is a close cousin to MRP. DRP works in the same manner by driving requirements from firm or forecasted signals. These demand signals can come from actual customers or forecasts. When business is seasonal or cyclical in nature, DRP can be helpful. Today's system tools are helpful in keeping watch over inventory at the distribution sites. Not so many years ago, this visibility was not always as available. In a DRP environment, when actual demand happens, the replenishment signal is combined with any additional forecasted requirements and a replenishment order is created. This allows seasonality to be built into the MPS to drive variation in stocks at different times of the year. DRP is a good system for complex environments with many products and varying demand cycles, but sometimes simple systems like demand pull are just as effective. The decision has to be made based on the complexity of the market and product demand. Many ERP systems today are set up for DRP functionality with the visibility of remote warehouse inventory for planning. These new ERP systems offer the flexibility of using both DRP and kanban in the same company.

Ultimately, in distribution, the master scheduler is helping to manage buffer inventory. The better job the master scheduler does, the better the customer service and the lower the cost. It comes down to how the demand plan or forecast is managed and how well inventory strategy (MTS, ATO, MTO, etc.) is maintained and linked to business strategy. Anyway you look at it, the master scheduler is in the middle of the process.

Web Added Value™

This book has free materials available for download from the Web Added Value™ Resource Center at www.jrosspub.com.

6

ORDER MANAGEMENT

Order management is a close associate of the master production scheduling (MPS) function in all companies. Order management team members are the people who respond to customer requests and promise deliveries to customers. In many organizations, order management reports to the sales team, not the MPS department, although there are exceptions to that. (In Chapter 17, you will read about The Raymond Corporation's approach, where order management does report to the MPS team.)

In order for order management to do the best job for both the customer and the company, it needs to be closely aligned with the scheduling process. This means the rules of engagement need to be clear and well understood and the available to promise (ATP) discussed in the last chapter well communicated. *Order management should have knowledge of and access to the ATP screens in the system.* The other key to this success is understanding the time fences and being able to translate the rules of engagement within these time frames. A typical time fence arrangement is depicted in Figure 6.1. This illustration will be used as the model to describe the normal characteristics of each time fence.

RULES OF ENGAGEMENT

Disciplines are an important part of a successful MPS process in any organization. Since order management is the group that usually makes the promises to customers, it is critical that these people, more than anybody else in the organization, understand the rules and why they are in place. *It is important to also note that any time the topic of rules of engagement is brought up, it must also be remembered that these rules are not established to say no to the cus-*

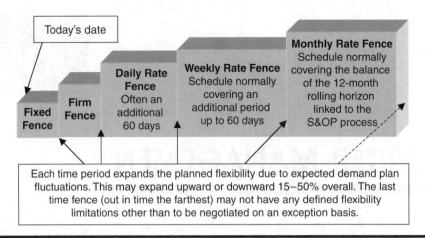

Each time period expands the planned flexibility due to expected demand plan fluctuations. This may expand upward or downward 15–50% overall. The last time fence (out in time the farthest) may not have any defined flexibility limitations other than to be negotiated on an exception basis.

Figure 6.1. Time Fence Norms and Flexibility Built into the Planning Horizon

tomers. Rules are established to acknowledge how much it costs each time you say yes. The time fences are set to help with the administration of reasonable rules. Each has a reason for existence, and each is meant to facilitate certain guidelines for customer promises as well as order policy and even how the MPS is maintained. Time fences are like speed zone signs. As we move from future planned orders into populated schedules, the master scheduler has to slow down and watch more carefully so as not to collide with a capacity constraint. The rules as illustrated in Figure 6.1 are the same as the fences used in the MPS maintenance and used to manage suppliers. The definitions can change and are related to the customer world of order management.

Fixed Fence Rules in Assemble to Order/Make to Order

In most businesses, the fixed fence is no more than one week. In many businesses with lean processes and flexible suppliers, the fixed fence can be as short as one or two days. The rules are normally simple — no changes. With that type of agreement, there is also a need to define the rules for how to *make changes* within the fixed period. Usually, changes inside this period have some impact on cost, and rules may define who is responsible for costs up to certain points of quantity or time requirements. When increases are made to the MPS inside the firm fence, it can mean increases in costs, especially if nothing is taken out of the schedule and material is being bought just in time. In some businesses, changes inside the firm fence can affect changeover time and thus machine or line downtime. Within the firm fence, the schedule is pretty committed in many businesses. Change policy should be well defined, with exceptions (which will happen) made only when clearly the best business decision. In businesses where the firm fence is shorter, such as 24 hours, the firm fence is less of a problem. It behooves any organization to increase flexibility, allowing the minimal need for firm fence rules beyond a few hours or days.

Determining the firm fence takes some knowledge in three areas:

1. **Market expectations** — If the firm fence is too long, it will affect market share and customer satisfaction. If the market requires that last-minute decisions be allowed as normal business, suppliers to that market obviously need to comply. The shorter the firm time fence is, generally the more competitive advantage is built into the process.

2. **Flexibility of manufacturing** — This is an area that needs constant attention for improvement. Every flexibility increase that can be made in the business will help competitive advantage. If, however, there are long setup times and changeover costs, it will tend to force the firm fence to be established with a longer time frame.

3. **Inventory strategy** — In an assemble-to-order environment, the firm fences can be quite flexible. This requires inventory somewhere in the supply chain and is a both favorable and unfavorable cost consideration, depending on the span of configurations and supply chain flexibility. Make to order generally requires a little longer firm fence. Make to stock is a different situation because the only solid fence rules are the rules governing ATP. If there is stock, it can be promised; if not, either swaps have to be arranged or the lead time is time to complete the next lot and more than just the normal shipping time.

Firm Fence Rules

Beyond the fixed fence, more flexibility would be expected in most supply chains. Where this comes into play in order entry or order management is when surprise demand spikes occur. This can affect all of the inventory strategies, even make to stock. If the normal demand for any one item is around 1,000 a month and an unforeseen or unplanned demand comes in for a one-time requirement for 10,000 pieces, obviously the master scheduler would make a decision as to where these would be placed. The right way to handle this type of exception is to have rules of engagement around the firm fence that describe parameters of flexibility. For example, there may be a handshake that allows the order management group to promise up to 15% above the current ATP outside the fixed fence. This might require the master scheduler to run planning bills of material in excess of the actual forecast and cut them as the bills of material approach the fixed fence without being exercised. Again, this would depend on the environment and the need for flexibility for competitive advantage. The difference between the fixed fence and the firm fence is flexibility. Rules define the amount of planned flexibility. Without this handshake, the master scheduler has to drive inventory to high-risk possibilities or run the risk of disappointing the sales team frequently as customers act in their natural unruly nature.

Daily Rate Beyond Firm Rules

Normally, order promising beyond the firm fence is less of an issue. Exceptions can include original equipment business gains, new product demand, and national accounts. In these few examples, demand is somewhat open out into the future. In the first time fence beyond the firm fence, guidelines are helpful if agreed to ahead of time. This allows the order management team to confidently hold discussions with the most important customers and give indications of company capability and readiness. Obviously, the farther the time frame is from today, the more important it is for order management to have discussions with the master scheduler. The typical business will set parameters to define the planned limits on flexibility at the daily rate fence beyond the firm period. While this can sound complicated, it does not have to be. A simple set of rules written on one page and easily accessible to order management is sufficient. Normally customers can wait longer for an answer on time frames beyond the firm time period, allowing for handshakes with the master scheduler. The real priority requirement to respond quickly is normally on orders required right away. These are in a different category.

It is not unusual for the flexibility rule at the daily rate fence to be set at 20% above (or below) the stated plan. The plan would be visible on a shared drive for all who need to see it. Order management would certainly be at the top of the list of people who need to know. When orders that break the agreed upon rules (in this case 20%) are received, order management needs to comply if at all possible. This means managing the situation to both the company's and customer's best interests.

Weekly and Monthly Rate Fence Rules

The farther down the lead time path the discussion goes, the less likely the forecasted rates are to be accurate. Again, rules can be easily defined for order management, including "immediate response" type of flexibility requests versus "consult factory" types of inquiries. This handshake makes everyone's job easier, both master scheduling and order management.

For the same reasons master schedulers move to weekly rates at some point within the 12-month rolling horizon, out at some point, the level of detail can be minimized again. This is done by simply moving from weekly rates by product family to monthly rates by product family. This keeps the maintenance work to a minimum and allows for easy alignment with the sales and operations planning process monthly. Again, flexibility is in open season.

These guidelines are defined as a general method. Your situation may vary somewhat, but these concepts are valid in all businesses. It all depends on market need, the flexibility of manufacturing, and the inventory strategy. *It is up to the master scheduler to make sure that the handshakes are established, communicated, visible, and maintained. The sales team has the responsibility to make sure the rules are competitive.* In many businesses, order management reports to the sales team organizationally. In these organizations, the two groups need to be very close. Many organizations are locating the order management team in the same room or office with master scheduling. *Some progressive organizations even have order management report directly to the master scheduling department.*

It is a simple world. Promises have to be within capability. Capability can be planned or not; sometimes it is easy and other times not, but it is a choice. Order management does not get to make that choice alone in most businesses. Top management has to step up to the responsibility of acknowledging and planning capacity — demonstrated capability. Master scheduling, sales, marketing, and order management all have to be part of the decision. When everybody is engaged with facts, the best decisions normally result.

OUTSOURCED PRODUCTS

It is becoming more frequent for businesses to expand their offerings by totally outsourcing some products from other original equipment (OE) manufacturers at the finished goods level. This has allowed companies to expand their product offerings without huge capital expense. In those companies, order management can experience additional complications, especially if the supplier is located offshore or far away. Again, the same rules come into play. If the strategy is make to stock, the rules are easier but the cost higher. If the expectation is make to order at the OE supplier, the rules become more important than ever if service fidelity is important. In that case, the master scheduler would work out the plan and rules of engagement with the master scheduler at the OE supplier and firm up the plans. Order management would deal with master scheduling as it would for other normal requirements, following the rules. In some businesses, order management works directly with the OE supplier. This works okay as long as the order management team approaches it with the same rigor and discipline as would the master scheduler.

ORDER MANAGEMENT SUPPORT FOR DISTRIBUTION CENTERS

Along with inventory strategy comes the needs for distribution. Touching product as few times as possible is always a consideration, but sometimes, because of lead time requirements and transportation costs, it becomes a good decision to warehouse product in strategic locations for quick response to customer need. In these businesses, it is important to manage these warehouses the same as you would any inventory — accurately and responsibly. Order management is done in different ways in various companies, but many handle orders directly placed on the distribution. Order entry may not necessarily be at the distribution facility (often it resides at a corporate location), but the emphasis should be directed toward "pulling" inventory from manufacturing to the distribution warehouses and on accountability at the distribution level for customer service. When inventory is "pushed" from the factory to distribution, there are often conflicts with space and priorities and negative effects from changeovers at the factory. The signals usually are the best if the pull is closest to the customer. Oftentimes, the warehouse develops a special relationship with local customers and has the best insight into their specific needs. Order management must link with this signal for replenishment regardless of where order management resides. This does not negate the rules of engagement. The rules should take this input into consideration.

Master scheduling would treat these orders from distribution just like any order. Normally, the inventory strategy at the plant would be make to order. Lead times would be built into the rules of engagement for distribution pull. Like any customer, exceptions would happen and a documented process to handle exceptions would exist.

The order management team is the conduit to the customer from master scheduling. This is a very important function and one that the master scheduler needs to have close ties to. This means constant if not frequent communication and a trusting relationship. Order management needs to be trusted not to make promises that cannot be met or are outside the handshake, and master scheduling needs to drive competitive advantage to make order management's job easier. This competitive advantage is in the form of shorter lead times and lower costs.

The master scheduler's job is broad and deep. It is a leadership function that spans several areas in the business. In the next chapter, we will get into the roles of the master scheduler and explain why these roles and this job are so important to the efficiency of a high-performance business.

This book has free materials available for download from the
Web Added Value™ Resource Center at www.jrosspub.com.

7

ROLES OF THE PLAYERS IN THE MASTER PRODUCTION SCHEDULING PROCESS

Before discussing the roles of the players in the master scheduling process, it is a good idea to list the process owners or leaders in this space. All role descriptions will be kept within the context of scheduling, management, and enforcement of the "rules of engagement" and order management. There are obviously other process owners and leaders associated with scheduling and planning, but this book will cover only those with direct connection to the master scheduling process. The processes connected to the rules and how they are enforced are by nature also connected closely to the master scheduler, and these positions will be covered. The center of it all is the master scheduler, however. What better place to start the discussion of roles.

ROLE OF THE MASTER SCHEDULER

There are varying organizational structures in manufacturing, but because most master schedulers are only involved in one manufacturing plant, roles will be discussed from that perspective. If the master scheduler in your organization has jurisdiction over several plants or multiple facilities, some translations will be required, but in general you will find the defined roles appropriate.

There are several important roles the master scheduler plays in a manufacturing organization. He or she is involved in many infrastructure events and even facilitates many of them. In this section, it should become apparent that the responsibilities of the master scheduler are more management than execution related. The master scheduler is one of the most influential positions in any well-managed business. Duties include starting the monthly cycle by updating the operations plan for top management in preparation for the sales and operations planning cycle.

Sales and Operations Planning

The master scheduler plays a primary role in the sales and operations planning (S&OP) process. *Although the operations plan is blessed by management, it is rare that top management actually develops the plan.* This preparation includes collecting and analyzing data and updating spreadsheets that later are distributed to top management for review in anticipation of the monthly S&OP review. During the S&OP process, the master scheduler normally is a messenger for top management measures and supporting data. *It is not uncommon for the master scheduler, because he or she is so close to the battlefield, to update the president or CEO on areas to question in the S&OP review. This update normally happens before the meeting in the CEO's office, which allows the top manager to be very prepared for the S&OP.* This is especially helpful for the top manager. It ensures he or she knows all the high-risk areas and accordingly asks the right questions to assess where risk exists in the manufacturing process and schedule. This is very important because the schedule equates to revenue. In most publicly held businesses, the CEO has made promises to the outside world considering financial success on a quarterly basis. Risks in this plan are top priority, or should be.

Following the monthly S&OP process, the master scheduler is responsible for developing the detailed schedule for production. That could be the exact plan developed and delivered to the S&OP review, but it could also be a modified plan based on decisions made in that meeting. The plan has to be revised and communicated to appropriate parties so that it will be implemented seamlessly.

Lastly, within the S&OP process, the master scheduler captures information for metric reporting on monthly top management plan accuracy in areas of operations planning and in many organizations, depending on the demand management function and organization structure, even in the demand accuracy areas.

Weekly Management Systems

- **Weekly performance review** — In the middle of the week, usually Tuesday at 1:00 P.M., the master scheduler often facilitates a weekly performance review meeting. This review is the accountability infrastructure for high performance or Class A ERP performance. Because the master scheduler is the main work organizer for the plant, this role is a natural. Any plant that does not have a weekly performance review would benefit significantly from it. Chapter 10 provides more definition detail on this important topic.

■ **Project review (weekly or monthly)** — Another important management system in high-performance organizations is the project review infrastructure. This is simply the management review of top projects empowered and engaged in moving the organization forward toward the business imperatives. In this position, the master scheduler would be a participant and generally not the facilitator. Because the master scheduler is required to be on top of all capacity changes, it is helpful for him or her to be a participant in the review.

■ **Clear-to-build** — The weekly clear-to-build process belongs to the master scheduler. It is in this process that the schedule commitment is affirmed once a week with the key players, production and procurement. Further discussion is provided in Chapter 10 on management systems.

Order Management

The master scheduler's activities also normally include communicating regularly with the order entry function, as well as demand-side personnel who are in need of special assistance outside the normal rules. Since the master scheduler's job is to keep production in synchronization with demand, there is no better group to be directly connected to. The demand review, which usually happens on the last Friday of the month at a minimum, is a main event between the master scheduling function and the demand side of the business. The master scheduler is typically the facilitator of this meeting (demand review was discussed in detail in Chapter 4). Most companies have a demand review weekly. This more frequent demand review is becoming more and more popular for good reason. Process continues to move faster and customer expectations continue to rise. This requires tighter linkage between promises and execution.

Communication with the demand-side people and processes does not stop with the demand review process. *It is not unusual for the master scheduler to communicate several times a day with order management, sales, or marketing when smooth communication flow exists.* In some organizations, the demand manager and the master scheduler actually sit together. This works very well in most environments.

The most typical master scheduler duties are:

- **Develop the operations plan** — Using up-to-date information, propose operations plan alternatives.
- **Data and metrics** — Keep the data and metrics for the top management metrics associated with the S&OP process.
- **Update the president/CEO for S&OP** — The master scheduler meets beforehand with the top manager to ensure specific questions are asked in the S&OP review meeting.
- **Facilitate the weekly performance review** — The master scheduler is generally the master of ceremonies for this important performance review.
- **Massage the weekly schedule** — As orders are received, the master scheduler schedules them accordingly.
- **Attend the weekly demand review** — The master scheduler is a main player in the weekly or monthly demand review process.
- **Process owner for master production schedule (MPS) metrics** — The master scheduler is responsible for root cause analysis and for driving action to eliminate variation in the MPS metrics.
- **Develop rules of engagement** — Every business needs rules to promise orders and keep promises in sync with execution. The master scheduler facilitates the development of these rules and also maintains the rules of engagement with the demand-side process owners.
- **Coordinate schedule changes** — The master scheduler updates and changes manufacturing schedules as required on a daily/hourly basis.
- **Assure MPS content** — The master scheduler is responsible for making sure that all business requirements are driven in the MPS. This might include aftermarket service requirements, demand review from products subcontracted out, scrap, etc.
- **Chair the weekly clear-to-build** — The master scheduler facilitates buy-in to the weekly schedule by production and procurement.
- **Determine planning bill of material design** — By using current knowledge and historical data, the master scheduler develops the planning bills of material structure for driving the "unknown" requirements.
- **Team member for new product introduction** — New product introduction is a major factor in MPS accuracy. The master scheduler must have representation on all new product introduction teams.
- **Involved in engineering change** — Effective dates of engineering changes are often coordinated by the master scheduler.
- **Master schedule maintenance** — The master scheduler maintains the daily massaging of the master schedule, including the creation of the time bucket size/design based on demand forecasts, S&OP, and agreed upon rules.

- **Lead time** — The master scheduler communicates lead times to the field, especially if lead times fluctuate due to inventory strategy.
- **Liaison between sales and operations** — The master scheduler facilitates proper communication between sales and operations.
- **Work closely with demand manager** — In organizations that have demand managers, the coordination between the master scheduler and this demand position is critical.

The role of the master scheduler indeed is big. The duties can be summarized into a few categories: communication between sales and operations, manage the S&OP process behind the scenes for management, develop the operations and master schedules, process linkage for schedules, coordination of scheduled capacities, maintenance of the 12-month rolling capacity buckets, and rules of engagement. This heavy load requires the position to be filled by capable people. Generally, this means product knowledge, leadership capabilities, and materials and operations experience. One of the best master schedulers I have worked with was actually the maintenance manager before going into sales management and then master scheduling. The fact that he had a photographic memory probably didn't hurt either! He is a nonego-driven type of person with the desire to factually describe everything using data and not emotions — a good combination.

ROLE OF TOP MANAGEMENT IN THE MASTER SCHEDULING PROCESS

No role is more important to ensure success of master scheduling than that of top management. Understanding the importance of schedule stability is more important than any other aspect of this process. In many organizations, top management is close to the customer and is quick to make promises that are contrary to internal agreements, which affects schedule stability. In high-performance organizations, top management understands the ramifications of such actions and gets the rules changed or tries to sell an alternative solution before taking the easy answer and corrupting the schedule. It is easy to say "whatever the customer wants to hear." It is high performance to anticipate this and have rules in place to facilitate it or to manage the customer to the best outcome for both parties when the wrong demand requirements were anticipated. Do not misinterpret this. *The customer has to be taken care of, but not always at unlimited costs to the company.* Good companies understand the market and have processes in place that match the market need. When the two are not in sync, costs are always too high or service too low. Top management must understand this and support it daily.

The following is a summary of top management roles as they relate to master scheduling:

- **Support the rules of engagement** — Top management must be party to the development of the rules governing order management and inventory strategy and help enforce them.
- **Facilitate the S&OP process** — Top management is responsible for facilitation of the S&OP process. Top management process owners must come prepared with answers for metric performance misses and not just rely on others to report at this meeting.
- **Drive root cause analysis** — If root cause or facts drive actions, the likelihood of business success is greatly improved. Top management can and must insist on this practice.
- **Schedule stability** — Top management can help maintain low costs by insisting on schedule stability (few changes in the short-term schedule). Stability duration requirements are different in each market and can be, for example, 2 hours, 24 hours, or 2 days. What matters most is that the organization has agreed on the fixed fence and (the majority of the time) adheres to it.

Obviously, top management has the most important role in the business. From a scheduling standpoint, that job is especially important. If the rules are current to market need and are enforced, master scheduling stands a much better chance of succeeding.

ROLE OF THE DEMAND MANAGER IN SUCCESSFUL MASTER SCHEDULING

The position of demand manager is closely associated with the master scheduler's role. These two positions are intermingled in many ways. *In effect, the demand manager is really the master scheduler equivalent on the demand side of the business.* The demand manager is normally the keeper of the order file and the person who massages the data inputs to the delivered demand plan otherwise known as the forecast. The demand manager is the conscience of the sales organization, making sure that all valid inputs are included in the forecast information and that risk is well recognized. The demand manager is a liaison between the sales people and master scheduling and has to understand both sides of the issues that arise in this relationship. This is not always an easy job, but if top management supports maintaining correct market-driven rules and enforcing them, the job can be fun and rewarding.

The duties of the demand manager as they pertain to master scheduling are as follows:

- **Coordinate the inputs for the forecast** — The demand manager must coordinate the collection of data from various sources for the demand plan. These sources can include sales people, customers, top management, marketing, promotions, etc.
- **Deliver the forecast and improve the accuracy** — The demand manager is the messenger and aid to the process owner for the demand side of the business on the demand plan. The work is often done mainly by the demand manager with inputs from many sources. It is almost impossible to have consistent accuracy in forecasts, especially in businesses that are trying new ideas for growth. Top management is the process owner for the accuracy of the data in high-performance businesses.
- **Rules of engagement** — The demand manager is normally the demand-side person who is actively involved with the master scheduler to keep the rules up to date and accurately followed.
- **Communicate frequently with master scheduling** — The demand manager and the master scheduler should communicate often to ensure proper coordination of the schedule.
- **Process ownership preparation** — The demand manager should be in constant communication with the vice president of sales and the vice president of marketing on issues regarding forecast accuracy.
- **Sales metrics** — The demand manager is often the one who calculates the performance for sales. This includes forecast accuracy, sales levels, customer retention, lost sales, etc.

The demand manager is an important internal partner with the master scheduler. Some organizations, especially smaller ones, do not have a demand planner. In these organizations, the master scheduler and functional sales management have to split the duties.

ROLE OF MATERIALS IN MASTER SCHEDULING

In many high-performance organizations, the materials organization is separate from the master scheduling organization and both report to the plant manager. Some of the more frequent possibilities are depicted in Figure 7.1.

In each of these organizational possibilities, the job duties of the master scheduler do not change. I have always found that if the master scheduling department reports to the materials department manager, either the master sched-

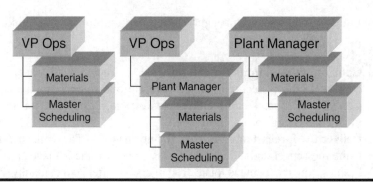

Figure 7.1.   Materials and Master Scheduling Organizational Possibilities

uler is still very openly active and carries authority or the real master scheduler is actually the materials manager, with both functions (materials and master scheduling) reporting to this person.

If separate in function from the master scheduling team, the materials manager is deeply involved with the success of the MPS and the master scheduler. The MPS is only as good as the inputs to the process. The materials manager, the one directing planner activity, is privy to a tremendous amount of valuable data on a daily basis. This might include suppliers in trouble with capacity, upcoming quality issues, late shipments, etc. *If the master scheduler is to have the best chance at schedule accuracy, all of this information needs to be captured and shared. The materials manager must create an environment where all planners feel obligated to help with the most accurate inputs and early warning systems are available.*

Materials manager duties that are important also include the following list of activities:

- **Manage slow-moving inventory** — The materials manager must be aware of slow-moving inventory and work down the list regularly with appropriate resources to eliminate it through a promotional sale, reworking it into something usable, or scrapping it out.
- **Manage obsolescence** — Budgeting for obsolescence should be part of the materials manager's job responsibility. No one wants to throw inventory away, but it happens, and the materials manager is the one who should police this effort.
- **Maintain order policy and lot-sizing decisions** — How the shop runs can be greatly impacted by order policy and lot size decisions. The

materials manager is the keeper of these policy decisions and should review them at least quarterly if not monthly.

■ **Maintain inventory strategy (make to stock, make to order, etc.)** — This may be the most important aspect of the materials manager's job responsibilities. It is extremely important to have lead time and cost in line with the market and competitive advantage. Often compromises are required. The materials manager is involved in all of these decisions, along with the master scheduler and other top managers in the business.

■ **Process ownership of the materials metrics** — Metrics such as percent of pickable orders released, orders released to suppliers with full lead time, and safety stock at goal percentage are examples of measures that the materials manager would typically own and report on. This ownership includes root cause analysis on misses as well as actions, names, and dates.

■ **Monitor disciplines in the execution of master scheduled requirements** — This is best done through the use of feedback metrics and reporting.

■ When the warehouse reports to the materials manager, additional responsibilities come to bear:

　□ **Inventory accuracy** — The materials manager, although not the process owner, will be a sponsor of inventory accuracy in raw material, work in process, and finished goods.

　□ **Where inventory is stored in the process (point of use or centralized storage)** — These decisions are made by the materials manager with handshakes from production.

　□ **Less than 24 hours dock to stock** — The receiving area needs to be as disciplined as any other area in manufacturing. Material needs to move immediately from receiving to the assigned area and show availability in the business system. The minimum acceptable time frame is *less than* 24 hours from dock to available for use. In some businesses, three hours is too long.

Other main players in support of the master scheduler include the production managers, engineering managers, and maintenance managers. Production obviously needs to be dedicated to adhering to the schedule as determined. If the schedule is not reasonable, the production manager needs to provide the appropriate feedback at the weekly clear-to-build process. It is not acceptable to change the schedule whenever the spirit moves the production manager to do so.

ROLE OF THE OPERATIONS MANAGER

Good operations managers are often the drivers of Class A ERP process and high-performance enterprise resource planning. They, of course, have much to gain from this in both ease of job and ultimate success. The best managers in this role are very supportive of the MPS and follow it as best they can. The clear-to-build process is operations' chance to buy into the following week's schedule. It is held at the end of each week, and the schedule is presented and approved by production (more on this appears in Chapter 10 on management systems). By working with the master scheduler and supporting the clear-to-build process on Friday and insisting that production accept its responsibility to scrutinize the schedules ahead of time, real and accurate schedules result. The front-line managers often do not step up to the "ownership" of the schedules unless they see a clear signal from management in this regard. They have a great deal of influence over the success of the plans. This influence includes equipment condition, changeover times, efficiency, knowledge of quality issues, vacation planning, and other production-related inputs to successful schedule execution. All of these types of process disciplines are signaled from the head of operations. What does he or she think is important? Everyone sees the signals.

The operations manager or vice president of operations is the pivotal role in the S&OP top management planning process as well. It is expected that the vice president of operations will participate in the S&OP by understanding the variation and discussing it appropriately in the top management planning meeting. In that meeting, he or she is the owner of the operations plan metric. This means the vice president of operations is responsible for communicating actions taken to eliminate root cause to process variations in the operations plan.

ROLE OF THE PRODUCTION SUPERVISOR OR LINE MANAGER

Probably the most influential people in the shop are the first-line supervisors or line managers. These people have direct access to the work being done each day and are also very involved in sequencing the activities of the workers. If they do not have respect for the MPS and the daily plan, it is unlikely that any master scheduling process will be successful. Line management must be on the same page as the rest of the organization. Goals and schedules must be shared. Getting these people on board requires the operations management to also be in sync with this need. When these supervisors are in tune with the shared schedule and are disciplined to follow it and help with the decision process, high levels of customer service can be accomplished.

Line supervisors must be party to the Friday schedule decisions at the clear-to-build event and influence changes when required during the review process. Once the schedules are in place, it is not up to the supervisor to make scheduling decisions. Obviously, if there was a mistake in scheduling, drop-ins affect schedule prioritization, or machine or tool problems develop, the supervisor must be in communication with the master scheduler to update the sequence on-the-fly.

Line supervisors can own their portion of the daily schedule and can report to the weekly performance review with actions and dates for continuous improvement to their schedule adherence.

ROLE OF THE MAINTENANCE MANAGER AS IT RELATES TO MASTER SCHEDULING

When machinery does not run reliably, it becomes very difficult to predict the daily or weekly capabilities accurately. Maintenance management owns the preventative maintenance process in a manufacturing organization. The measurement of success is the uptime consistently and reliably enjoyed. This is measured by the total hours available or running. Any time a machine is not "ready and available to run" is subtracted from the total hours available, even if no work is scheduled for the machine. This means that setup time, machine breakdown and repair time, and preventative maintenance and maintenance time are counted as "badness" against the metric.

Preventative maintenance has been proven time and time again as valuable and must not take second priority to production. If there is not enough capacity to both produce the schedule and do preventative maintenance, production management must increase capacity to meet this need. This can be done through overtime, hiring, adding shifts, or adding machinery.

Maintenance time requirements such as preventative maintenance and machine rebuilding should be planned into the MPS by the master scheduler. To not do so is negligent and will eventually come back to haunt the master scheduler and the production team.

ROLE OF FINANCE IN MASTER SCHEDULING EXCELLENCE

Finance is the conscience of the organization. It is also the risk management planner for future periods. All projects require the involvement of finance to verify savings. After all, if the savings are misunderstood, the priorities might be wrong in assigning resource. Finance also plays an important role in the

S&OP process. Its guidance is critical in making mix and business planning decisions regarding product emphasis. Just a couple of generations ago, the finance group was focused on reporting what happened yesterday. Today's finance group has a much broader role. It is also expected to understand what is likely to happen tomorrow through understanding previous behaviors and results and translating current trends.

ROLE OF THE READER OF THIS BOOK

As the reader of this book, it is your job to assess the balance between the existing state of your organization and the descriptions of process and accountability in this book. Top management does not always see the same level of inefficiency that people in the trenches do (obviously it can work the other way around also). If you are one of the people from the trenches, it may be appropriate to bring some of the points to the attention of those in top management. If they are worth their salt, they will appreciate the feedback. If everyone carries their weight, top management's job becomes easier, as does everyone's. Good master scheduling is essential for high-performance manufacturing. Master scheduling success happens because of good discipline.

Web
Added
Value™

This book has free materials available for download from the
Web Added Value™ Resource Center at www.jrosspub.com.

8

SALES AND OPERATIONS PLANNING

The sales and operations planning (S&OP) process, or the SIOP (sales, inventory, and operations planning) as it is called in some companies, is one of the most exciting and most talked about topics in business today. The interesting aspect about this is that the S&OP process is not new; in fact, some might argue that it is an old process. There is just a lot of new interest in it, and for good reason — it pays off. Organizations are working not only internally but also with their suppliers and in some cases even their customers to help each other's top management planning process.

Virtually all high-performance organizations do some form of an S&OP process regularly. In these organizations, the S&OP is a monthly top management planning meeting where metrics and performance are reviewed and adjustments made based on recommendations concluded from data collection and analysis done in preparation for this review. The keys to success are preparation and good data mining in advance of the decision process and top management support or engagement. Top management should run these meetings.

AGENDA COMPONENTS OF THE S&OP PROCESS

This first activity is to evaluate performance for the current period (last 30 days). This agenda item is followed by an analysis of the 30-60-90-day time frames going forward. Beyond 90 days, on an exception basis, the rest of the 12-month rolling horizon is reviewed.

The components of the S&OP are as follows:

1. Review actions from last S&OP meeting.
2. Review of last 30 days — The accuracy metrics by product family are reviewed.
   a. Financial plan accuracy by product family — The top management financial manager (usually the chief financial officer) communicates root cause analysis of any process variation and actions to improve upcoming accuracy.
   b. Forecast plan accuracy by product family — The top management demand-side manager (usually either the vice president of sales or the vice president of marketing) communicates root cause analysis of any process variation and actions to improve upcoming accuracy.
   c. Production plan accuracy by product family — The top operations management manager (usually the vice president of operations) communicates root cause analysis of any process variation and actions to improve upcoming accuracy.
3. Review of 30- to 60-day plan expectations — Risks and/or changes since the last monthly plan are reviewed in detail.
   a. Financial plan risks
   b. Sales forecast risks
   c. Production risks
4. The 90- to 120-day plan expectations — On an exception-only basis, this time frame is detailed.
   a. New product introduction is discussed for risks of plan accuracy.
   b. Production shifts to alternative sources such as moving product lines from one internal plant to another, migration to offshore sites, supply chain risks, etc.
   c. Promotions, shows, customer actions, etc. are reviewed on an exception-only basis.
   d. Review/verification of normal cyclicality/seasonality is done.
   e. Anticipated currency exchange issues affecting the plan accuracy are discussed.
5. The balance of the 12-month rolling schedule is quickly reviewed for anomalies or changes expected.
   a. Any new information is communicated.
   b. Changes from the previous plan are reviewed.

PARTICIPANT ROLES IN THE S&OP

There can be a few variations in participation and roles depending on the organizational structure and size of a business. For the purpose of this chapter, two views will be used as examples: a larger multiplant environment and a small company.

In a larger organization, often product managers will be responsible for overall product success, including creation of demand, product cost, and overall profitability. It is a highly responsible job, and even though product managers are not always considered top management, they really are in the solid core of people influencing the planning process. Larger organizations also usually have a demand manager within either the sales or marketing leg of the business. In some organizations, the demand manager reports to the vice president of sales, in some the vice president of marketing, and in yet others the vice president of sales *and* marketing; again, it depends on the size and scope of an organization.

In this example, we will assume that responsibility for both sales and marketing functions is held by one vice president, as well as the normal expectation in most businesses of having a chief executive officer, chief financial officer, vice president of operations, and of course a master scheduler. Figure 8.1 depicts the S&OP participants under these assumptions. Note that in this example there is no demand manager; the forecasting process would be executed through the product managers. A demand manager is a strong asset, but some companies decide not to organize that way. In bigger companies where demand is complex and covers numerous product families, a demand manager almost becomes a necessity to coordinate the various plans into one overall plan. In some smaller companies, the demand manager takes the place of product managers. All of these various approaches have been used successfully. There is no need to shy away from any of them. The approach needs to fit your budget, top management commitment, and company size.

In smaller organizations, there are generally fewer top managers, and each wears more hats. Because of this, participation in the S&OP will engage a smaller group, but the duties are all still executed. In smaller businesses, there may not be a product manager for each product family, but there still could be a demand manager position. The demand manager normally reports to the ranking demand-side manager and collects data from the field and consolidates input information to develop and maintain the demand forecast. Figure 8.2 shows the typical S&OP participation in such an organization.

Very small organizations do not usually have demand managers. This does not stop the process, but the duties have to be picked up by the sales or

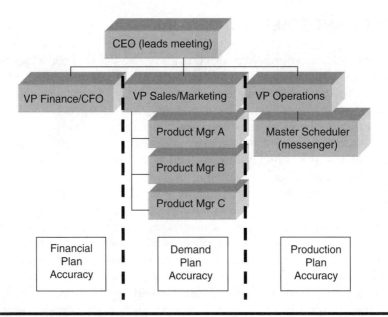

Figure 8.1. Typical Participation in the S&OP in a Larger Business

marketing administration. In some organizations, the responsibilities are covered by having the clerical pieces picked up by an administrative assistant and the analysis done directly by the vice president of sales and marketing. None of these scenarios is better than another; it simply depends on the complexity

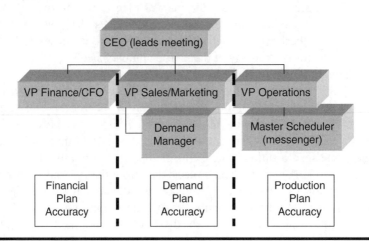

Figure 8.2. Typical Participation in the S&OP in a Smaller Business

of the market, number of products, appetite of the vice president of sales, and size of the organization.

A QUICK WORD ON ALTERNATIVE LABELS FOR S&OP

In some organizations, the acronym S&OP has an additional letter. This is prevalent enough to include the acronym SIOP in this text. SIOP is an abbreviation for sales, inventory, and operations planning and is synonymous with the S&OP process. Similarly, PSI (production, sales, and inventory) is another acronym used in this space. In my experience, organizations that have picked up the PSI version and abbreviation tend to share a lower evolution of S&OP DNA and are often missing some of the important elements of planning and accountability. There is nothing scientific in this observation, and very little has been written about the ancestry of the various terms beyond S&OP, the most often used moniker for the popular top management planning process. Most of the time, when companies refer to these additional process acronyms, they are speaking of the top management planning process.

ROLES IN THE S&OP MEETING

The monthly S&OP meeting in a high-performance organization will have a very predictable agenda and roles of the players. First of all, the meeting belongs to the CEO or president, whoever is in charge regularly day to day. This monthly meeting is a main element of the Class A ERP management system, so accountability needs to be a top priority. Accountability does not mean bloodletting is required. Instead, it means that process owners come into the meeting prepared, with their homework done, and present the facts, including actions to offset process variation. Accountability is best ensured if there is someone to play the role of asking the tough questions. That role resides with the chair of this meeting, the president or CEO. Each of the other plans has a process owner, and it is the role of each of these owners to propose updates or changes to their plans to help eliminate plan inaccuracies. The duties of each person in the meeting are described in the following outline. Some alternative participants are also listed. Every organization will be a little different in its choice of structure. There is no right or wrong organization chart as long as the right process owners are accountable for the accuracy and facts relating to their plans.

Duties of Each Process Owner
in the Monthly S&OP Meeting

1. President/chief executive officer
   a. Maintains the meeting schedule 12 months in advance (usually done by always holding it on the first Tuesday of every month, or second workday of every month, or some related predictable schedule).
   b. Leads the meeting.
   c. Makes sure the meeting is not preempted by some other priority.
   d. Ensures the priority of attendance so process owners all show up regularly.
   e. Monitors and facilitates a consistent agenda.
   f. Asks the tough questions as metric performance and actions are presented for each plan (financial plan accuracy, demand forecast accuracy, operations plan accuracy).
   g. Minutes are normally published from his or her office.
2. Vice president of finance/chief financial officer
   a. Prepares the financial spreadsheets for the S&OP.
   b. Presents metric performance numbers for financial plan accuracy, by product family, past 30 days, and sometimes the current quarter and/ or yearly plan.
   c. Shows trends of product family financial plan accuracy, month to month, for at least the last six months.
   d. Shows delta between plan of record and the last product family financial forecasts.
   e. Presents actions to close any accuracy gaps.
   f. Updates and shares product family financial plans going forward, 30-60-90 days and balance of 12-month rolling plan.
   g. Highlights any changes affecting promises to stakeholders.
   h. Keeps the financial records in synchronization with the master production schedule.
3. Vice president of sales and/or marketing
   a. Presents metric performance numbers for demand plan accuracy, by product family, for the past 30 days.
   b. Shows trends of product family demand plan accuracy, month to month, for at least the last six months.
   c. Shows delta between demand plan of record and the last product family forecasts.
   d. Presents actions to close any accuracy gaps.
   e. Updates and shares product family demand plans going forward, 30-60-90 days and balance of 12-month rolling plan.

    f. Highlights any changes affecting forecasts, such as promotions, large shows, phase-outs, new product introductions, etc.

    g. Answers the questions in the meeting concerning the forecast accuracy.

4. Vice president of operations

    a. Presents metric performance numbers for the operations plan accuracy, by product family, for the past 30 days.

    b. Shows trends of product family operations plan accuracy, month to month, for at least the last six months.

    c. Shows delta between plan of record and the last product family operations forecasts.

    d. Presents actions to close any accuracy gaps.

    e. Updates and shares product family operations plans going forward, 30-60-90 days and balance of 12-month rolling plan.

    f. Highlights any changes affecting plans, such as supplier risks, production shifts from one facility to another, and/or any other unusual circumstances.

    g. Answers the questions in the meeting concerning the operations plan accuracy.

5. Master scheduler

    a. Prepares the demand and operations plan spreadsheets and performance metric performance, by product family.

    b. Distributes the demand and operations plan spreadsheet to the S&OP participants prior to the meeting (as far in advance as possible).

    c. Meets with the CEO/president prior to the S&OP meeting to make sure he or she knows all of the issues and can ask all of the right questions.

    d. Is not at the meeting to answer operations questions, although the master scheduler can be of assistance to both the operations and demand plan process owners in the meeting by providing data and facts.

6. Other possible S&OP meeting attendees:

    a. President's administrative assistant

      i. Takes notes during the meeting, including any assignments or agreements for actions.

      ii. Publishes the minutes.

    b. Demand manager (if the organization has one)

      i. Prepares the demand spreadsheet instead of the master scheduler.

      ii. Ensures that the demand and operations spreadsheets are in the same format and are in complete synchronization.

      iii. Provides facts and data in regard to the demand plan.

   c. Product managers (if the organization has them)
      i. Provide forecasts for their product families.
      ii. Present actions affecting customer behavior.
      iii. Answer the tough questions regarding plan accuracy of their product families.
      iv. In many organizations, these people share full accountability with the vice president of sales/marketing for plan accuracy. They normally report to the demand side of the organization.
   d. Vice president of engineering
      i. The vice president of engineering is often a valuable attendee at the S&OP process meeting. This process owner is usually well served to understand the priority and issues around risks in upcoming schedules as well as longer term, especially as new products affect these schedules. New product introduction typically is one of the biggest risk management opportunities in the planning horizon. Engineering obviously has a big role in successful plan accuracy of new product introductions.

TIMING FOR THE S&OP PROCESS AND INTEGRATION WITH THE FINANCIALS

The S&OP process needs to happen as quickly each month as possible. In many high-performance organizations, this means that the meeting is always held on the second or third workday of the month. This makes last month's information very fresh and allows decisions to result in changes that can affect the current month while there is still time to make a significant difference. Some organizations have decided to hold it later in the month. The normal reason given is they are waiting for the books to be closed. The best organizations do not worry about the books being closed as it relates to the S&OP process. In fact, there is an advantage in the books not being closed. After all, this meeting is about plans and planning accuracy. When the demand and operations plans are accurate, the financial plans are often equally accurate. When these plans are not accurate, there is a great learning opportunity. If you are in the financial department and are wondering how the results can possibly be reviewed at the S&OP if it is done prior to the books being closed, you might want to consider recalibrating your paradigm for this model.

The financial results should be directly tied to the demand and operations plans. If, for example, the operations spreadsheet planned certain costs by product family, producing more or less units would affect the costing process results. This can be estimated quite closely if the fixed and variable costs are separated. (Fixed costs are things like heat, property taxes, some portion of utilities, some indirect labor support activities, etc.) This modeling can get as sophisticated or complex as you want, but the simpler models seem to work the best as they are more easily understood in the S&OP discussions. When plans are not made because volumes were too low to offset burden costs, this can be easily explained. This is often more palatable than complex models that increase fixed costs automatically below certain volumes and have to be explained every time they come up in the S&OP discussion.

Applying the Financial Impact to the S&OP Spreadsheet (See Table 8.1)

Assumptions:

1. Product line A is a make-to-order (MTO) product family.
2. In product family A, the cost per unit/cost of sales including labor, burden, material, and other overhead is approximately $1,000.
3. In product family A, the "per unit" average revenue based on planned mix is approximately $1,895.

It is helpful and quite appropriate to calculate this model for each of the 12 rolling months. Of course, the "actual" fields would only be populated for past months. By building a spreadsheet that includes costs and revenue estimates per product family, the job of planning becomes much easier. This again raises the discussion about product families. One rule that is not as often applied to product family designations is price and/or margin levels. Price and/or margin can create the need for additional product families when bands in price or cost are too broad. Like all topic ranges, if the range is too broad, the family relationship can become less meaningful. *The general rule is to keep product families under 10, with 6 as an ideal number.* Having said this, it also needs

Table 8.1.   Applying the Financial Impact to the S&OP Spreadsheet

	January Example			
	Actual Units	**Planned Units**	**Actual Revenue**	**Planned Revenue**
MTO demand	100	115		
MTO operations	115	115		$217,925
Gross margin	$102,925	$102,925		
Gross margin = (115 × $1,895) − (115 × $1,000) or $217,925 − $115,000 or $102,925				

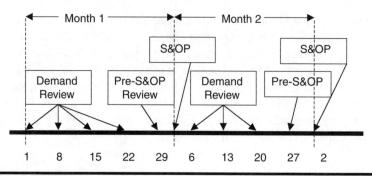

Figure 8.3. Timing of the S&OP Cycle: Example of Possible Friday Calendar Dates in Two Monthly Cycles

to be said that there are lots of examples where businesses have built successful S&OP processes with more product families than the preferred six. Just use some common sense to keep it as small as logically possible.

Some of the more obvious rules that most people are aware of to develop product families include: shared business constraints, shared components, shared markets, and common manufacturing sites. There can be others.

The timing for the monthly cycle of Class A–related activities in S&OP preparation includes the demand review, the pre-S&OP, and the S&OP meeting itself. The demand review normally happens every Friday except for the last Friday of the month. On the last Friday, the demand review is replaced by the more detailed and encompassing pre-S&OP meeting. The pre-S&OP covers all the demand review topics and includes a more in-depth review of both performance and risk analysis going forward.

The scheduling of the S&OP process is important (see Figure 8.3). It needs to be timely and repeatable/predictable. If it is done at the same time each month, no one will miss it with the excuse that they did not get the communication. If it is done very close to the beginning of the month, this perfect timing allows decisions to affect the current month.

AGENDA FOR THE PRE-S&OP

The pre-S&OP is a meeting held just prior to the S&OP meeting. Since it would not be acceptable to show up at the S&OP without root cause analysis completed and proposals for continued improvement prepared, it makes sense to have a gathering of process owners just prior to the main top management meeting to determine issues and analysis actions as required. A company CEO deserves to see recommendations and action plans at the S&OP, and that is exactly what gets accomplished at the pre-S&OP.

The players with required attendance at the pre-S&OP are:

1. Product managers (if the organization has them)
   Product manager alternative: Demand manager
2. Demand manager (if the organization has one)
   Demand manager alternative: Product manager or vice president of sales and/or marketing
3. Sales manager(s)
4. Master scheduler
5. Plant managers often attend, although in large companies this might be via conference call

The discretionary attendees are:

1. Vice president of operations
2. Vice president of finance/CFO
3. Vice president of sales and marketing (if they are well represented by a demand manager or product manager)

The master scheduler is normally the facilitator at this meeting. The agenda is always the same, which allows for players to come prepared with actions and proposals to present. Decisions will be made and handshakes will happen, but each player should anticipate these needs and come prepared for them. Some of the kinds of decisions that have to be made or proposals that have to be confirmed include inventory plan changes, forecast changes, capacity increases or decreases, changes to marketing promotions due to availability of product, etc. These decisions can become pretty involved, and it is helpful to the business to get the data and facts out in the open for full disclosure. The better the job that is done at the pre-S&OP, the faster and more efficient the S&OP process will be. Think of the pre-S&OP as the analysis gathering and proposal development meeting, and think of the S&OP follow-up as the decision-making meeting where the proposals are either authorized or adjusted by the top management team. High-performance organizations get very good at preparing for the S&OP by anticipating questions that might be asked and having the answers either ready or presented in anticipation of the questions. The master scheduler again is the main facilitator and the person in charge of the quality of the data and analysis going into this meeting. This is not because the master scheduler does it all, but because the master scheduler is the one who makes sure the right questions get asked and sends out pertinent data ahead of time.

The agenda for the pre-S&OP is basically the same agenda as for the S&OP, just at a different level of preparedness. In the pre-S&OP, there are often unanswered questions. In some instances, this meeting is the first time all of these players have been in a room together to discuss these topics since the last meeting a month ago. At the end of the day, however, the topics are the same as the final top management planning meeting that follows this one by a few days. Those review steps in sequence are: review the financials, review the sales forecast, and review the operations plan. The sequence of detail would be:

1. Review last 30 days of performance
   a. Demand plan
   b. Operations plan
2. Review changes required going forward in current 30 days
   a. Demand plan
   b. Operations plan
3. Review the 60-90-120-day forecast for risk management and new issues
   a. Demand plan
   b. Operations plan
4. Briefly discuss any issues longer term. This is a much shorter topic here than in the S&OP. The main focus in this meeting is the 30- through 120-day horizon.

THE SPREADSHEETS FOR BOTH THE PRE-S&OP AND THE S&OP PROCESSES

The format for the typical S&OP is pretty consistent business to business (see Table 8.2) There is a separate spreadsheet for each product family, and the history has to go at least 30 days prior to the current review period.

Additionally, some organizations use a step chart or what is sometimes called a waterfall spreadsheet to see the progression of plans over time (see

Table 8.2. Typical S&OP Spreadsheet

			Product Family A (4 Months of a 13-Month Spreadsheet)						
	Act.	Plan	Perf.	Act.	Plan	Act.	Plan	Act.	Plan
Demand plan	115	100	85%		105		109		115
Operations plan	105	100	95%		100		110		115
Financial plan	$109	$105	100%		$105		$109		$113

**Table 8.3. Step Chart in the S&OP Format
(Full Spreadsheet Would Run 12 Months)**

	Jan	Feb	Mar	Apr	May	Jun	Jul	Aug	Sep	Oct
Feb plan	10	10	12	12	14	14	12	10	10	10
Mar plan		10	12	12	12	14	14	12	12	10
Apr plan			10	12	14	14	14	14	10	10
May plan				10	14	16	14	12	12	10
Jun plan					12	18	14	12	12	10

Table 8.3). This can be done by keeping the 12-month rolling plan intact and adding the new plan to the bottom of the spreadsheet each month.

Note that for the sake of space, this spreadsheet does not show the full 12-month rolling horizon required by good S&OP process and Class A ERP criteria. If this waterfall spreadsheet is used regularly, top management can gain additional insight into changes in plans through the year. In this application, for example, one could look at August or May and see how these plans have changed month to month. Sometimes this is helpful in understanding the dynamics of the planning process. Seeing these shifts is a good reminder that the supply chain is constantly adjusting to the forecast updates, and often this can lead to additional costs. It is too easy to forget this fact without the information in front of the right people.

SUMMARY: KEYS TO THE SUCCESS
OF THE S&OP PROCESS

The S&OP process is very important to the success of any manufacturing business. There are a few keys to remember for high-performance payback that have been proven in many businesses in the last 15 years:

1. Meetings must be scheduled 12 months in advance. Use known scheduling equations, such as every Friday for demand reviews, the last Friday in the month for the pre-S&OP, and the second workday of the month for the S&OP.
2. The meeting should not be preempted for any reason short of an unscheduled catastrophic surprise. When a scheduled S&OP management system event falls on a holiday, the next workday should be automatically scheduled. No exceptions. If it is scheduled as, for example, the third workday, this problem does not exist.

3. No excuses for nonattendance are acceptable. Of course, all rules are made to be broken, and short of sickness, serious personal issues, vacation, or an exception-only customer conflict, all top managers are to attend the S&OP. Some companies have allowed sales people to miss any time they want. This behavior does not favor high performance. When key players are missing, the meeting does not promote proper accountability or achieve the true handshake between the demand and supply sides of the organization. The demand review has a little different roster, but the same rules apply. On those rare exceptions when there is an absence, it is expected that a fully authorized backup will attend in the missing player's place. Authorized backup means that this person has the full authority of the process owner and can make decisions accordingly.

4. The spreadsheet needs to include all the plans: financial, demand, and operations. With the exception of the financial plan, the master scheduler is typically the facilitator of the format, plans of record, and metrics, especially for the demand and operations plans. The financial plans are typically prepared and presented by the CFO.

5. Keep the agenda consistent, meeting to meeting. The president or CEO should require the team to stick to a time frame and keep the team on task.

6. The first thing to review in the meeting is the performance number (percentage of accuracy). When the performance number is reviewed first, there is a sense for good or bad that is not always apparent when just talking about the actual volume numbers, actual versus plan. For example, sometimes 460 out of 500 may sound acceptable, but when there is an acknowledgment that 95% is the threshold of acceptability, 92% (460 divided by 500) is no longer acceptable performance and is quickly recognizable as such. The organization needs to be tuned in to 95% minimums for schedule and data accuracy performance. Anything less is just unacceptable and requires root cause analysis and actions/names/dates for eliminating these root causes.

7. The vice-presidential process owners need to answer the tough questions in the S&OP process review meeting. Organizations that completely delegate the accountability to lower levels, with vice presidents sitting back and blistering the second-tier people in the S&OP meeting, do not reach the same levels of proficiency in the business. When vice-president-level process owners take the time and interest on a daily basis to understand the risks and are willing to coach the management on them, there is always a business benefit. This culture can only be

encouraged by the CEO or president. It cannot be done from the grassroots.

8. The president/CEO has to have a very clear interest in not only the financial numbers and performance but also the underpinnings of unit performance. The top manager must be willing to get into enough detail to understand risk management of the plans. Things like production transfers from one facility to another and new product introductions are of special concern. These issues should be clear and visible well ahead of the actual implementation of change. This way, the right questions can be asked in advance and proper investigations and actions initiated by the time questions are asked in the S&OP meeting.

9. The master scheduler should have access to the top manager (CEO or president in charge of the S&OP meeting) prior to the S&OP meeting. The vice president of operations, to whom the master scheduler normally reports, must also understand that this exposure and preparation for the top manager (CEO/president) to the master scheduler's information will make the process much more valuable to the business. To appease protocol, sometimes the vice president of operations goes to this prep meeting with the master scheduler and top manager. *Because of this required relationship with the top manager and because vice-president process owners should be answering their own process-related questions during the review, it works best if the master scheduler is not frequently asked for answers in the S&OP meeting. Instead, the master scheduler should be thought of as the messenger.* This ensures that the master scheduler is keeping the right issues and tough questions on the table. This risk assessment knowledge, best held by the master scheduler, is some of the best information the business owns. Expose it.

10. Actions should always result from discussions during the S&OP meeting. The CEO's administrative assistant is an excellent person to take the notes during the S&OP meeting and to publish the minutes from the meeting. In the business where I grew up, the president and CEO's administrative assistant was masterful at capturing the essence of the action requests and handshakes from the meeting. This report was always a catalyst for action. It is a well-know fact that, too often, good intention is overridden by day-to-day obstacles and procrastination. Try to avoid this.

11. Review last meeting actions first thing before starting the normal S&OP agenda items.

12. Accuracy metrics concerning the top management plans should be posted in the factory and office areas in a visible place. The names of

the vice-president process owners should be listed with the performance percentages.

13. Celebrate successes when appropriate.

Businesses that follow these rules and put honest effort into reaching genuine handshake agreements between the demand- and supply-side operations enjoy great benefit. The S&OP process is a key element of the Class A ERP certification and, even more importantly, the payback from it. In Chapter 9, the focus will turn to the ERP business model and the metrics that govern good manufacturing disciplines today. Master scheduling is the first operations planning process below the S&OP processes and is the heartbeat of ERP application. The metrics are the window into performance understanding and required accountability.

This book has free materials available for download from the
Web Added Value™ Resource Center at www.jrosspub.com.

9

MEASURING THE MASTER PRODUCTION SCHEDULING PROCESS

Measurements are the feedback loop to both performance and improvement in general. The master production schedule (MPS) is one of the most important processes in an enterprise resource planning (ERP) business system and therefore should have robust measurements to drive that performance and improvement. Several metrics can help with this goal. The first one most organizations use is the weekly schedule attainment measure. This metric is done by locking the schedule (for measurement purposes only) at the end of the week. During the following week, changes are monitored and documented to understand accuracy and root cause.

The following are some reasonable rules to incorporate into your measurement process:

1. The metric is facilitated by the percent of complete orders that are completed in the week they were scheduled according to the latest weekly schedule (plan of record from the end of last week).
2. There is a separate schedule (and metric) for each product line in the business.
3. Each schedule is locked at the end of each week for the following week (for measurement purposes only; schedule will actually be revised as necessary).
4. The consolidated plant or business metric is the accumulated percentage of orders completed within the week scheduled. No averaging of product line performance averages.

5. If rule 4 above is followed, weighting of values within the metric is not necessary. If there are some product lines with numerous small orders and others with few large orders, it doesn't make any difference!

6. The planned schedule and actual comparison is to the detailed configuration stockkeeping unit (SKU) and order quantity. If the order is for 1,346 pieces, the measure is to 1,346 pieces. A missed order is a missed order. It can be missed because of time, quality, or quantity.

7. If the orders scheduled for completion are all completed with full quantities, the plan is met and the measurement performance is 100%.

8. Any orders that are not completed within the week to full planned quantities are "misses" in the metric.

9. Products that do not make it through normal quality checks in the assembly process by the end of the week and create uncompleted orders are counted as misses in the master scheduling performance metric.

10. Products that need rework, if reworked within the scheduled time, are schedule hits, not misses. Note that it is not the intention of any process to require rework, but other measurements will pick up the first-time yield opportunities. The master scheduling metric is simply designed to focus on schedule accuracy. When metrics are designed to pick up too many factors, root cause analysis and process ownership are both negatively affected.

11. The metric is calculated just as any percentage. The numerator is the number of orders completed to the original plan of record compared to the denominator, which is the number of orders planned (see Figure 9.1).

MPS METRIC WHEN INVENTORY STRATEGY HAS LESS THAN A ONE-WEEK "FIXED" FENCE

There is a little twist to the metric when building short-cycle products, especially in an assemble-to-order or make-to-order environment. In these environments, the plan of record when locked on Friday the prior week can be mostly unscheduled orders. Take, for example, a saw blade company in the Boston

$$\frac{\text{Number of completed orders (right quantity)}}{\substack{\text{Number of orders planned} \\ \text{in most recent plan of record}}} = \text{Percent MPS accuracy}$$

Figure 9.1. MPS Accuracy

area. It receives orders for big saw blades, sometimes up to 42 inches, that are made per customer application for cutting hard materials like brick, concrete, and even metal. The company makes the blades specifically to the application from scratch and ships in less than 72 hours. In this organization, the master scheduler has no idea what will be built and shipped two days from now, to say nothing about what will happen at the end of the week. The master schedule consists of planned capacity to be held, and as orders are received, this capacity plan is consumed. The metric is executed a little differently in this environment. The metric is still comparison of actual to plan of record, but the plan of record is the first time the order is scheduled. All orders that are completed in the week they were originally scheduled are considered hits (not misses).

The master scheduling metric is the feedback mechanism to make sure the weekly master schedule is as accurate as possible. High-performance minimums require at least 95% performance in this calculation. It takes both a good plan and good execution to achieve a high-performance MPS process.

There are other appropriate metrics in manufacturing that link to the MPS process. Two important ones are the procurement process metric and the schedule stability measure. Yet another is the materials planning metric. All of these metrics are commonly found in a Class A ERP management system.

PROCUREMENT PROCESS METRIC

A high-performance procurement process measurement is typically calculated as the percent of complete orders or releases from purchase orders that arrive on the day or hour (depending on the schedule interval) the parts were required. For this metric, the definition of "the day or hour the parts were required" will be found in the ERP business system. The required date/time is the latest requirement date/time in the system at the time the material was received. That means the date/time that has been affected by any changes since the order was released. The definition of "complete" means exact quantity to the latest order or release and 100% usable parts — no rejects. Any reject makes the order a miss for the day. If an order is "shorted" by manufacturing or the supplier, it is a miss even if planning adjusts the order quantity (which planning should do anyway in most cases). If there are multiple SKUs on one purchase order, the full order with all of the SKUs is the definition of the requirement. This metric is at a higher standard than many procurement process measurements, but the lessons learned from this will be very beneficial to the business. Remember that the objective is not "95% in boxes"; the objective is high performance. Many times, the supplier is not the problem at all; the schedule stability or the MPS maintenance/accuracy is. Without this metric in place and visible, it is more difficult for the purchasing professionals to get their point across concerning process deficiencies.

The traditional supplier metric of measuring the integrity of the last promise from the customer is a reasonable lower level metric and often is a good source of data for the procurement people in manufacturing. Alone, this metric does not meet the correct metric requirements for high-performance MPS and procurement linkage and system integration and does not get into enough detail to consistently point to root cause.

MATERIALS METRICS

Metrics for materials control are also very logical. Class A criteria for planning accuracy allow for some choices here. The following are a few suggestions:

1. Percent of purchase orders released without full lead time as defined in the lead time field on the item master in the business system
2. Percent of 100% pickable assemblies released on time to assembly
3. Percent of 100% pickable orders released to the warehouse on time
4. Percent of schedule changes within the fixed period fence (usually 48 hours but can be up to one week)
5. Inventory turns — raw, work in process, and finished goods
6. Percent of components finished on time into stock or available on time

Each of these metrics brings some special goodness to the table. Each organization needs to determine what the biggest need is in terms of the required drivers of action. If your business is an assembly shop and it is currently common to have orders released for pick without all the components available, the percent pickable metric is very valuable. If your business has myriad changes within the fixed period that cause excessive cost from changeovers and material movement, then the schedule change metric is recommended. High-performance or Class A ERP criteria require at least two metrics in this area of focus. There are no bad metrics in this list, and implementing all of them can be a risk-free compromise.

IMPLEMENTING METRICS

Implementing metrics requires a process owner for each one, a management system for reporting progress, and a standard reporting format that includes performance, trends, root cause analysis, and actions/names/dates for improving the performance. There are two types of metrics in high-performance or Class A ERP process. Bob Shearer, of The Raymond Corporation, one of the first and best master schedulers I have worked with, many years ago determined that the normal Class A ERP metrics always appropriate in every business are "barometric" measures — they tell you if a storm front is coming. These metrics, however, do not necessarily tell you what the cause is. These indicators only point you in the right direction. The barometric measures should be maintained even when the performance level is sustained for many months. The reason these metrics continue to exist is for audit purposes — understanding things are still in control from that perspective. An example of an MPS barometric metric would be the number of complete orders finished in the week scheduled by the Friday prior MPS plan (the weekly schedule record is frozen for measurement purposes only). By measuring the deviations from plan and understanding the root cause for the changes, much can be gained in the business.

Another level of metrics required in every business is called the diagnostic measures. These measurements can change from business to business and should. They are driven from root cause seeking and focusing on areas of weakness in a specific situation. Environment, skills, process variation — these things are all different in different companies and therefore will require different metrics. That is why they are referred to as diagnostic measures. In the case of the MPS, there might be a specific customer that is especially "unruly." Diagnostic measures may be developed around this specific customer to look for underlying trends and information. This would be an example of a diagnostic-type measure (see Figure 9.2).

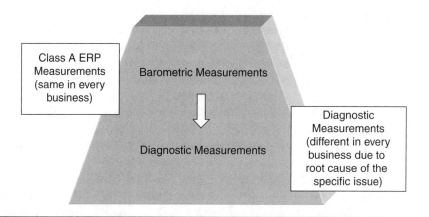

Figure 9.2. Barometric Measures Versus Diagnostic Measures

STARTING THE METRIC PROCESS

Some businesses measure every process, and this is expected and taken for granted. In these companies, it has become part of the culture. In other businesses, as metrics are added to the deck, resistance can be seen. Performance measurement in high-performance organizations is not "something they do"; it is "the way they think."

Typically, there is not a great desire to measure performance throughout the business in the beginning stages of implementation. As is human nature, people often see this change as a threat to their security or even a violation of their honor or trust and think, "Why do you want to measure my performance anyway?" or "Do you not trust me or do you think I am not doing appropriate work?" This is the first phase of performance measurement. There are three.

Human Acceptance Phases of Performance Measurement

In my experience, there are stages people go through when first being introduced to performance metrics. They are usually similar to the following:

- **Stage 1** — "It's not me you need to measure; it's him/her" (points finger at another employee). This is probably the denial stage.
- **Stage 2** — "Okay, I guess you [management] are not going to forget this new idea" (with sort of a disgusted tone). "What was it you wanted measured anyway?" This is the acknowledgment stage — it is real. "It's going to happen, and there isn't much we [the employees] are going to do about it."
- **Stage 3** — "I have been measuring this and I found that there are some other diagnostic indicators that will help in the elimination of the root cause of the variation." This is the exciting stage — the stage of discovery.

Obviously, education plays a part in this transition, as does practice. Management must send a clear and consistent message that the measures are about process and not people. This is done through both actions and words. Management's signal cannot be delegated to the staff. The employees watch and know what management thinks is important! People always do what the norm of expectation is.

The most successful metric process start-ups involve education and communication. Normally, management starts the process, and once the metrics are confirmed from a standpoint of data and process ownership, they are made visible. Visibility has many benefits, not the least of which is awareness. Metrics create visibility.

10

MANAGEMENT SYSTEMS

Management systems are those events and processes that are in place solely for the sustainability or control of repeatable processes. Management systems normally are exercised in various time frames (daily, weekly, monthly, etc.). Metrics in themselves do not drive change; it is the management system alongside the metrics which does that. High-performance companies, in every case, with *no exceptions*, have robust management systems. Some of the management system events found in high performance or Class A ERP performance are:

- Monthly sales and operations planning (S&OP) process
- Weekly or monthly project review
- Weekly demand review
- Weekly (metric) performance review
- Weekly clear-to-build process
- Daily schedule review or walk-around
- Daily documentation updates and reviews

Figure 10.1 illustrates the relationship between two of the management systems and master scheduling. S&OP, covered in Chapter 8, is shown, as is project management.

In the monthly management system called S&OP, there is a focus on overall accountability and follow-up. It is in this space that management defines the priorities and makes sure these objectives are accomplished. The S&OP process, covered in detail in Chapter 8, is a major element of policy decision making and risk management. Project management is also a nonnegotiable

Figure 10.1. Monthly Management System Elements in Class A ERP S&OP Project Management Review

component of continuous improvement manufacturing. That would leave little to add here were it not for the need to tie this all together. It is communicating the *idea* of a management system that really gets people in the organization to understand what management is trying to accomplish by it.

The monthly management system is about setting policy and making decisions about the direction and tactics month to month. Top management owns the monthly space. The weekly management system is a bit more detailed and incorporates most of the rest of the organization. That would include all process owners, middle and line managers, and supervisors. The weekly management system requirements typically are:

- Weekly performance review
- Clear-to-build
- Weekly project review
- Weekly demand review (covered in Chapter 4)

WEEKLY PERFORMANCE REVIEW

The weekly performance review is one of the most powerful elements of the management system. With the exception of the S&OP process, there is no process more influential within a Class A ERP process. One of the main Class A ERP rules followed for the weekly performance review is that this meeting is never preempted. The weekly performance review has top priority for time management.

The weekly performance review meeting is best held toward the beginning of the week to review the performance of the preceding week. Each process owner reports the performance and actions driven from this process to the rest of the operations management staff. Tuesday, for many, has proven to be the best day for this review, as Monday is too early in the week to be fully prepared with analysis of last week's performance and misses. Tuesday afternoon has been found to be the last point in the week where the week is still young but there is adequate time to prepare for the review. The meeting should be held at a consistent time each week, whichever day is chosen. This eliminates the possibility of issues such as "I did not get the e-mail" or "I didn't know when the meeting was."

At a minimum, metrics normally reported at the weekly performance review in high-performance manufacturing organizations include but are not limited to:

- Master production schedule adherence
- Schedule stability
- First-time quality
- Bill of material accuracy
- Inventory accuracy
- Item master accuracy (this can be more than one metric)
- Procurement process accuracy
- Shop floor accuracy
- Customer service (this also can be several measurements)

The weekly performance review process is an integral part of a robust management system that allows management to be involved in the follow-up of process control on a regular and predictable schedule. It helps establish accountability for process ownership. Without it, process ownership has little or no real meaning to the organization and will not be effective. In a high-performance organization, as metrics are added to the deck, these metrics normally are added to the weekly performance review. It becomes "how you run the business" in terms of process ownership and follow-up.

By using this type of form for reporting progress and performance, the organization becomes consistent and effective at problem solving and continuous improvement. It should be used consistently for all performance metrics. This means that as metrics are included in data analysis, it would make sense to have these metrics reported at the weekly performance review meeting.

The Agenda for the Weekly Performance Review Meeting

The weekly performance review meeting should be predictable and repeatable. The agenda should be simple. Each process owner reports his or her progress for last week. Progress includes all elements on the quad chart in Figure 10.2.

Top Left of Quad Chart

In this example, the process owner for inventory accuracy is reporting process performance. This format would be used by all process owners. In the top left quadrant, daily performance is reported along with the weekly overall performance summary. Most high-performance metrics are expressed in percentage points. Performance is to be reported as a daily number and as a weekly totaled number.

Top Right of Quad Chart

The trend chart is displayed in the top right quadrant of Figure 10.2. This is important in determining if the process is affecting performance positively. It is not clear if progress is being accomplished successfully if only performance for the current period is reported. The trend chart should cover a minimum of five to six weeks of performance data. The trend graph reported in this quadrant

| MPS Accuracy | Process Owner: Joe Green | Date: xx October xxxx |

	Orders	Hits	%
Mon	35	34	97%
Tue	42	42	100%
Wed	37	37	100%
Thu	37	35	95%
Fri	45	45	100%
Total	196	193	99%

Weekly Trend

Pareto of root causes

Actions	Names	Dates
1. XXXXXX	XXX	XXXX
2. XXXXXX	XXX	XXXX
3. XXXXXX	XXX	XXXX
4. XXXXXX	XXX	XXXX
5. XXXXXX	XXX	XXXX

Figure 10.2. Quad Chart Performance Reporting Format

is typically weekly data points only. This creates a smoothing effect that can give a better and truer trend summary.

Bottom Left of Quad Chart

A Pareto chart showing the reasons for misses will allow the organization to focus energy and resource on the most important barriers to successful accuracy. This is especially important early in process focus and measurement when process is affected by the most sources. The best-performing organizations focus resource on the worst root causes, allowing the biggest return for their investment.

Bottom Right of Quad Chart

Process improvement comes only by driving change through actions. By institutionalizing and reporting the action process, follow-up is much more systematic and predictable. The root cause bar at the left side of the Pareto chart should have actions easily linked on the right side of the quad chart (Figure 10.2).

It is common to struggle with real root cause. It is not always easy to uncover root cause, but that must be the objective. The old rule of asking "why" five times helps to achieve this. A simple root cause rule is to ask why until the answer is actionable. Management has to play a role in this if it is to become a required behavior and a predictable component of the management system.

CLEAR-TO-BUILD

The clear-to-build process was reviewed in previous chapters, but because it is an important management system event and is a master scheduling event, it is also included in this chapter.

The clear-to-build process consists mainly of a commitment from production and procurement managers to the master production schedule (MPS) for the coming week. The master scheduler distributes the new MPS late in the week. In most businesses, this is on either Thursday evening or Friday morning. On Friday, the key players are expected to either confirm their commitment and confidence at a gathering or, in more mature, experienced, and well-master-scheduled environments, no meeting is held and the process owners of the plan execution only report back if they require adjustments prior to being able to commit.

In most organizations with a good robust MPS process, the clear-to-build happens just once a week, but in some cases it can effectively happen more than one time a week. In one organization that turns raw and component inventory approximately 40 times a year, the schedule is fixed (stable) only about three days into the future. This company's markets are connected directly to retail sales and need to be highly flexible if demand changes quickly. In this environment, the clear-to-build is especially critical and is therefore done *every day*. In the daily mode, at 1:00 P.M. the master scheduler meets with the purchasing and production managers and assesses the status of inventory for the following day's build. The objective is to have confirmation of everything required for tomorrow by the 1:00 P.M. meeting.

WEEKLY PROJECT REVIEW

In many organizations, high-level management personnel review projects at least once a week. It totally depends on how many projects are in process, how important the outcomes are to management (high-profile projects are reviewed more frequently), and what the impact or risks are. The important concept here is the predictability and repeatability of the project review process. Once a project is assigned, there needs to be clear understanding of when and how often it is to be reported and reviewed.

DAILY MANAGEMENT SYSTEM EVENTS

It has been said many times that if you get the day right, the week will go well. If the week goes well, you don't have to think about the month. This is probably a little simplified in terms of the real world, but the spirit of this statement is correct. Since many businesses still have an "end-of-the-month crunch," there is probably something to be learned from daily management systems. The following is a basic short list of daily management system elements:

- Daily schedule alignment
- Daily walk-around
- Visible factory boards
- Shift change communication

Daily Schedule Alignment

In high-performance enterprise resource planning (ERP) applications, it is expected that all information in the system will be as current as there is knowledge in the business. Every time a situation changes, the system should reflect that change (within reason). That means there could be almost no past due requirements in the ERP business system at any one time. There is plenty of process variation in the world; all processes have it. That being the case, and given the importance of keeping the system up to date, alignment needs to happen often, at least every day.

The alignment is as simple as an assessment of production schedule adherence from the previous day and adjusting accordingly. In some businesses, there is very little extra capacity. This if often true in high-volume repetitive manufacturing. When the schedule is missed, time and capacity are lost; they cannot be recovered. The schedule needs to reflect that change. By the same token, if the business has some flexibility, misses from the prior day can sometimes be made up by working some overtime, doubling up on resource, etc. Either way, the schedule should reflect the latest expectations. When the MPS is updated, the material requirements planning tool can align signals to the supply chain accordingly, and both inventory and shortages are minimized.

It makes sense to have alignment in the system happen at least once a week, but in most high-performance businesses it is done every day or even every hour as required. This is probably the best approach. When schedules are kept accurate and realistic, the result is minimal system noise. The schedule should be reviewed daily by the production managers and materials management as well as master scheduling. This often happens in a daily communication first thing each morning. Some organizations do this through phone calls, while others have a short update meeting. During this meeting, the focus is on alignment — how many were completed yesterday on the line and how many in the sequence we expect to be able to finish today. This is discussed on each line, agreement is reached, and the master schedule is updated accordingly. In high-performance organizations, this can go quite fast. If there are lots of problems, this process can get slowed down. In many manufacturing plants, daily schedule white boards are bolted to the machine lines. In the daily schedule alignment process, the boards are updated each day with yesterday's performance and any changes to the schedule. The boards should be big enough to be easily read when walking by. The master scheduler collects and confirms any changes and the new schedule is updated into the system.

Variation can come in both directions. Poor schedules or poor execution that is difficult to schedule accurately can result in schedule variation to both the high and low sides. Also keep in mind that in a high-performance manufacturing organization, this may only involve 5% of the business on any one day.

Daily Walk-Around

The daily walk-around is exactly what it sounds like. Management reviews schedule adherence each day by walking around the facility to each line supervisor in his or her respective department to get an understanding of the progress made the day prior and what the expectations are for the current day. This visibility helps in several ways:

- It helps enforce the belief that schedules need to be accurate.
- It keeps management in the loop in terms of any variation developing.
- It allows management to acknowledge good problem-solving behavior.
- It gets management in the loop to help with more difficult issues when appropriate.

One example always comes to mind when I talk about the daily walk-around. While working in China, I was doing the final Class A audit in a plant in Shanghai. In this plant, the daily walk-around consisted of four people (the plant manager, the production manager, the master scheduler, and the engineering manager) meeting each morning to review the schedules in the facility. These four managers would meet in the factory each morning at the same time to tour the facility and talk to each line supervisor. In this particular facility, there were no middle managers. All of the line supervisors reported to the production manager. Line supervisors were team leaders who worked on the line as team members most of the time.

The reports I observed that morning during the audit were interesting. I met the other daily walk-around attendees in the designated aisle at 8:00 A.M., the usual time. We walked to the first work cell. The supervisor stopped his work and greeted us. The board showed no misses from the previous day, and the supervisor informed us that he did not expect to miss any schedules on the current day. All good, but not too exciting — yet. The same thing was reported on the second line, but when we reached line three, I started to perk up. Line three had missed several units the night before. Its performance was posted at 92%, which is unacceptable in this organization. As reported, the following happened.

The third shift had trouble with a machining center. A motor had burned up on the machine and production was halted. The third-shift supervisor went into maintenance, got a new motor that was stocked there, and assembled it himself. In most businesses I have visited, this would be quite impressive in itself, but it didn't stop there.

The machine was restarted and production was able to minimize loss to the schedule. After production was back on track, the supervisor then went back into maintenance to *fix* the problem. He pulled the records for the machine and found that the scheduled preventative maintenance was to be done in about two weeks. This preventative maintenance would have included taking the motor apart and checking and cleaning it. All of this would have probably prevented the unscheduled machine-down situation experienced that night. On that same shift, the preventative maintenance schedule was changed to a shorter interval by the supervisor, with communication left for the maintenance manager.

I do not see many supervisors who are that well trained and empowered to fix their own equipment and who also understand that this was not the real fix to the problem. This operator had been trained to ask the question "How do I stop this from ever happening again?" This is clearly the spirit of high performance and continuous improvement.

The walk-around is designed to keep management in the loop and to keep accountability at the line level. It is not expected that, given safety and training concerns, every team would be able to do its own machine repairs, but understanding root cause can be a big help to any and every process owner.

The daily walk-around is helpful. Several processes can be reviewed during this tour, such as:

- Schedule adherence
- Inventory accuracy
- Housekeeping and workplace organization
- Progress on recent projects

With the busy meeting schedules of most plant managers, this is a favorite time for many. When things are running as they should, it also gives the plant manager another good opportunity to show appreciation by thanking people and acknowledging good performance in public.

Visible Factory Performance Boards

To keep the schedule well communicated and understood, most high-performance factories are not relying on paper schedules hung by the machines. Instead, these well-managed companies are using large white boards with the daily schedule information posted for all to easily read (see Figure 10.3).

While this tool is used in many organizations, variations to the visible factory performance board layout are acceptable and common. For example, many organizations also put safety and first-time quality performance data on the board. Some organizations choose to do schedule checks more than once a day. One client I am currently working with chooses to have schedules checked twice a shift. On the day shift, the schedule checks happen at 10:00 A.M. and 2:00 P.M. The benefit of this is the communication value to ensure everyone is on the same page when they start the day. It also helps to underline the importance of making schedules as planned.

Schedule ___/___/___			Schedule ___/___/___			Perf
Product	Quantity		Product	Quantity		MTD
	Plan	Act		Plan	Act	
XX	23		XX	78		
XX	111		XX	99		
XX	5		XX	455		
XX	589		XX	4		
XX	45		XX	135		
Total	773			771		
Performance XX%			Performance XX%			
Open Issues			Open Issues			

Figure 10.3.   Visible Factory Board

Shift Change Communication

Each time a shift change happens, there is an opportunity to lose continuity in production. Each shift means new people come in to replace the people who are presently working on the schedule. In some businesses, it means that the processes and speeds will vary. In high-performance businesses, this is not the case. Much effort is devoted to keeping operators educated and trained to understand what the knobs on the machine really do and to setting up standard procedures to be followed by all operators. At one manufacturer in Baltimore, the manufacturing organization measures all changes to process. Every time there is a change from the "standard" process settings required to bring product into spec, a miss occurs on the line's metric. The metric is reported at the weekly performance review meeting and receives a lot of attention. This helps to focus operator awareness on the root cause. Many things have been learned from this. It is a good practice to adopt.

Shift communication is difficult. The people from the leaving shift are ready to go home; hanging around to answer questions is not always the most effective means of communication. Instead, many high-performance companies will establish formal communication processes. Formal and visible communication boards on the line are helpful. Having a diary-type log in the cell is also helpful. Good performance criteria suggest that the process be designed for your environment, but that it be consistent throughout your factory. This is just common sense and worth the extra time to establish that good habit.

OTHER MANAGEMENT SYSTEMS IN THE BUSINESS

Management systems are everywhere in high-performance businesses. In businesses where these process elements are second nature and woven into the fiber of the organization, there is no discussion about management systems. They just happen. For example, whenever there is a process change, the documentation is automatically changed because it just makes sense. If a metric is changed or a new metric is born out of need, the natural first thing to do is include it in the weekly performance review system. All projects are naturally included in some type of project review. In these companies, no decision is required to do this — it is obvious.

This describes the Class A ERP thinking and environment. Predictable process only comes from a combination of two things: a good plan and good execution. One without the other spells failure at one level or another. There are no exceptions. Thought and planning on management system elements can foster a culture where high-performance thinking happens naturally. All well-done and beneficial processes such as Class A ERP, lean, or Six Sigma are built around good robust management systems. Without these accountability infrastructures, organizations would not be as efficient. Management systems are just good process.

This book has free materials available for download from the
Web Added Value™ Resource Center at www.jrosspub.com.

11

INTEGRATION WITH MATERIAL REQUIREMENTS PLANNING

Master scheduling is so integrated with the rest of the business that it is often confused with other business system elements. Making it more confusing to the untrained eye, master scheduling also uses calculations and methodology similar to the material requirements planning (MRP) engine. They are about as alike in principle and practice, however, as night and day. That said, many companies do not treat the master schedule with the stature it deserves and the stature that is required from high-performance enterprise resource planning.

MRP is really the execution process in the schedule planning methodology. Master scheduling is the prerequisite planning process. Figure 11.1 shows that the real decisions are made in the master production scheduling (MPS) process, not in the MRP or materials planning phase of planning. In businesses where the MPS discipline does not exist, these important decisions are left to the planners. It is almost impossible for a planner who only sees a portion of the business to plan for the entire capacity picture. Most businesses have more than one planner, and they are not always crisply separated by product. In many of these businesses, some products have overlap in common processes, shared machinery, or components. Only when a business is very small does it normally work well for the material planner to also play the role of master scheduler. One such business manufactures saw blades for cutting concrete and blacktop. The business is small enough to have one planner do all the supply chain ordering

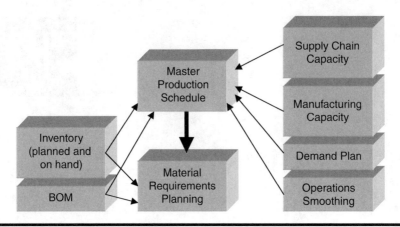

Figure 11.1.   MPS/MRP Hierarchy

and the master scheduling. In effect, this master scheduler wears two hats. This business is small enough to get away with it. Most businesses will not fall into that category.

PLANNER LOADS

As a reference number, 800 to 1,200 part numbers per planner is a reasonable number, although many businesses will vary from that due to simplicity or complexity. It is just a number to use as a reference. In one business I worked with, there were 70,000 part numbers regularly used in production and 150,000 bill of material records. This kind of complex environment requires the separation of planning (MPS) and execution of plans (MRP). Master scheduling plans the risk analysis, and planning sends the signals to the supply chain according to the plan established by the master schedulers. Most business have between 1,000 and 50,000 regularly used part numbers. Some have more, but the rules do not change with volume, and most of these businesses find that 80% of the volume is within 20 to 30% of the part numbers. All of this means that most businesses require both a master scheduling function and a material planning process.

PLANNER ETIQUETTE

The rules for the material planners in relation to the MPS are as follows:

1. The master scheduler has to be informed of any changes in capacity immediately so that the MPS can be updated accordingly. Even if no changes to the schedule are required, the communication should still happen because the master scheduler makes risk assessments every day based on such information. This might include a supplier that has machine-down problems, a supplier with not enough capacity, weather-related problems, ships arriving late into dock, etc.

2. The planners are the first line to the supply chain on a daily basis in most businesses. This often makes the material planners the best source of real lead time data regarding supplier performance. The planners should be either updating the lead time field in the item master or making sure this is done in a timely manner by reporting changes to the proper field owner, which often is the purchasing agent. By keeping this field accurate to the latest information, the chances of on-time manufacturing are increased.

3. Planners also have access to real-time data from the factory floor. If they are the first in materials/scheduling to learn something important, such as scrap issues, rejects, or machine-down situations, the planners need to communicate this immediately. The master scheduler may or may not decide to change the MPS, but this knowledge is important.

4. Occasionally, time standards (labor standards or cost standards) will be wrong on items being manufactured. Occasionally, the planner may know this information when the master scheduler does not. This can easily be the case with subassemblies not directly scheduled by the MPS. Although normally a production responsibility, if it is not done in a timely manner, planners should take the initiative to have inaccurate standards changed through the proper channels to keep data as accurate as possible.

5. When suppliers are generally misbehaving or not particularly trustworthy in their promises and feedback, the planner is often the first to realize it. This needs to be communicated to both the purchasing department and the master scheduler. Action must be taken as the situation dictates. When good performance measurements are in place, this is easy to detect and react to. The metrics and the clear-to-build process can act as early warning systems.

6. The warehouse operation can become a liability as well. The planners are not the only people to observe difficulties, but can be a critical part of the early warning. When the warehouse is filled up or capacity problems of any kind arise, the planners need to be proactive in raising a red flag.

7. Inventory accuracy trends in the warehouse or anywhere else in the operation are also critical to the smooth MPS execution. When problems

start to develop and are not being taken care of quickly, the planners may also want to raise a red flag.

As you can see, the planners are not only concerned with the proper planning of inventory and component availability, but also are an early-warning system for many areas affecting the MPS. The planning department can help serve as a conscience to the manufacturing supply chain mechanism.

INTEGRATION OF THE SYSTEMS

MRP and the MPS are always fully integrated in high-performance enterprise resource planning (ERP) applications. This does not mean that they are always or necessarily required to be within the same software application. The MPS is a capacity planning and customer satisfaction scheduling system at a high level. In some organizations, this schedule is done separately on spreadsheets and fed into the ERP system after the smoothing and maintenance are done by the master scheduler. At this point, you may be thinking that it would be crazy not to want this important schedule to be totally integrated within the ERP business system. Of course this is the ultimate goal, but many older systems were not implemented with MPS modules or have been mutilated by well-intentioned but misinformed implementers, and the functionality is simply not a short-term reality. In these environments, the MPS can still effectively be done off-system and downloaded into MPR each time maintenance is done, which often may be once or twice a day or even more in some businesses. The key to success is having the download happen easily and as automatically as possible. It is not difficult today to have Excel or Lotus spreadsheets directly linked to the requirement fields in MRP. Most of the newer ERP business systems have spreadsheet integration capability built into them in numerous places.

Delaying good MPS practices because your software does not have the MPS module is a bad decision. Excel or Lotus can be a great bridge, and the lessons learned will all apply to the new system if and when it arrives.

The *ultimate* goal is always to have the MPS module fully integrated with the MRP engine within the ERP business system. This allows changes to be seen immediately by the planners and problems quickly communicated up the line to the master scheduler as well. All of the popular bigger company ERP software suppliers have decent MPS modules, and when implemented as the systems were designed, they work quite well. To name these systems here would be to forget some. You know which ones they are. There are about five to eight that are really well known. There are many others beyond the top five to eight as well, but some of the lesser known software packages do not do justice to the MPS process and need to be scrutinized prior to implementing them. It has been my experience that the greatest diversity of functionality is in the group of software packages that are offered to run on PCs or smaller servers. Some are quite good, but many are dangerously oversold. I came across a small consumer goods company recently that was running its manufacturing process on a very well-known and popular PC small-business financial package. This package is a great financial tool; I use it in my business, but it is not designed as a manufacturing planning tool. The result was the need for continuous human involvement and manual intervention. This is not an example of good process!

THE MPS LINK

The data that need to be linked to the MRP engine are twofold. First are the actual requirements inside the firm and fixed fences that will drive the detailed schedule short term. Second is the information that comes from the top management planning process called sales and operations planning. The short-term details should be updated frequently as required. Actual demand coming from customers affects these short-term data minute by minute in some businesses. The longer term data, planned orders in the future, do not need to be downloaded as frequently. These future requirements might only need to be realigned once a month. In this case, exceptions would be done as required, with normal synchronizing done only periodically within the S&OP process. If the master schedule is planned properly, there should be no concerns about too frequent synchronization. Remember that good master schedulers do not make changes within the fixed fence unless absolutely necessary. Measure these changes, in both frequency and severity. The metrics along with good management systems help watch for these trends and drive actions accordingly.

Web Added Value™

This book has free materials available for download from the Web Added Value™ Resource Center at www.jrosspub.com.

12

FORECAST REQUIREMENTS

No topic generates more controversy than forecasting. Production people think it is the answer to all their troubles, and sales people think it is crazy to expect any accuracy from this process. The result in some organizations, unfortunately, is that production or materials professionals do the company forecasting without the demand-side people involved, and the sales people just plug in dollar amounts for the budgeting process. As the saying goes, "This is no way to run a railroad — or a manufacturing organization!"

Forecasting, as defined in Chapter 4, is a demand-side process and needs to belong to the demand-side experts. Unfortunately, perfection in accuracy is a dream. Moderate accuracy can be achieved, but to think that it will be sustainable without a lot of work is optimistic at best. The problem comes from companies that are using new services and trying new products. High-performance companies, by nature, are the most difficult to predict because of special causes introduced into the processes every day. These come from new markets, new products, and new services offered to entice customer loyalty and business. In this chapter, we will "poke at" the idea of how good the forecast needs to be. This is not an exact science, and markets and manufacturing processes can differ in many respects. Because of this, there will be no magic answer, but ideas and thought may help the process design of forecasting in your company.

WHO OWNS THE FORECAST?

In high-performance companies, the sales and/or marketing team owns the forecast and its accuracy. This means that the accuracy should be the concern of the demand-side team, and steps should be taken to continuously improve the accuracy. When the forecast is grossly overstated or understated, obvious questions need answers, such as "What can we learn from this?" and "What will we do differently next time to stop this from happening again?" Top management needs to ask the questions, and ownership needs to be obvious and in the right corner.

WHY SHOULDN'T PRODUCTION GENERATE THE FORECAST?

Production has no firsthand knowledge of the marketing efforts or customer tendencies, at least not to the extent the sales people have exposure. Production people also do not talk to the market every day, as the demand-side people do. This advantage of the sales people gives a lot more credibility to the forecast if the right amount of energy is put into the process. When production generates the forecast, the only input can be history. History only works when the company and markets are not changing, not a characteristic of high performance.

In many organizations, there is not enough top management support for forecast accuracy to be a focus. Sometimes this is because the president grew up in sales, or other times it is just because the top manager has not thought about the impact to the business. Top managers, like others, sometimes fall into the trap that "there is no such thing as an accurate forecast." While this is true, there is always room for more knowledge. That is the benefit that the forecast work really yields — more market understanding and knowledge. What top manager wouldn't be in support of that? The real work must be done by the demand side of the business if there is to be real improvement. Without that commitment, it is simply statistical data. One business I visited recently had hired a statistician from Harvard to work the forecast information. The accuracy had been improving, but only slightly. Problems were being driven from changes to the product offerings and new programs offered. To include these properly, the statistician required that these factors be added to the equations. Who was able to do that for him? You guessed it — the demand-side team. At the end of the day, these are the people who need to be accountable for the gains and performance of customer behavior.

HOW GOOD DOES THE FORECAST NEED TO BE?

The forecast needs to be as accurate as the process will allow. This means that it may not be all that accurate. The areas described in Chapter 4 on demand planning influences cited business priorities, marketing steps, sales knowledge, and history as the inputs. This activity, done professionally and with reasonable effort, can usually bring results worth the time spent to understand it. Sometimes markets just do not yield predictability. One business I have done work with is in the defense sector and supplies the U.S. Army and Marine Corps with equipment. As you can imagine, it has had difficulty forecasting the timing of various wars and international conflicts. Still, 75% of its work is in more normal market demands and could be forecasted.

The truth is that some markets are not predictable. Examples include new markets not yet established, completely revamped existing markets, and markets where something extraordinary is going on, such as changing technology or the competition changing strategies. This can work both ways and result in an over forecast or an under forecast situation.

Sometimes it is an advantage to the business to understand that the market is totally unpredictable. In these cases, top management must be willing to step up to the risks associated with this knowledge vulnerability. At the sales and operations planning process, top management must acknowledge that one of two strategies must be adopted: either risk customer service by keeping inventories low to an expected lower demand picture or run inventory higher and risk cost problems and overstocking materials. When the problem is ignored and left up to the materials organization, the accountability for forecasting and risk management falls back on that organization. This is not always the best method and is rarely the method of choice in high-performance organizations.

With the right level of learning and aggressive improvement incentives, reasonable markets should be predictable to an 85 to 90% accuracy level in units per product family. This is not to suggest that forecasts should be highly accurate at the stockkeeping unit (SKU) level. In most of the companies I work with, we do not measure SKU forecast accuracy. Review Chapter 4 on demand planning influences for more insight into this topic. This is normally measured by averaging the product line accuracy performance numbers.

HOW DETAILED SHOULD THE FORECAST BE?

The forecast should not be in any more detail than is reasonable for accuracy. SKU is too much detail in almost every market, and dollars is way too general and not much help to materials or scheduling. In most manufacturing businesses, there are generally about 6 to 10 product families. It is difficult to make this generalization because some companies are much more diversified than others. One business I worked with had 16 families initially; another had even more. At the company with 16 families, products from one family to another had very little in common in terms of parts or capacity requirements. Each product family can have wide variation in components, especially if they share most of the same capacity constraints.

One often missed element of product family design is the inventory strategy. Inventory strategy is important to separate because of the differences in scheduling technique required to properly plan and schedule units in production. A particular family of units may have one family for make to stock (MTS) and another for make to order (MTO). In each case, the forecast must be done in the same families. This means that the demand-side team needs to have a forecast for product family "A-MTS" and one for product family "B-MTO." Each would be measured for accuracy.

Other designations for product family division would include capacity constraints, shared components, and cost/price. If units within the same family are not in the same market (one is inexpensive and one very costly), sometimes it makes sense to divide these product families a little further. The danger is always taking the level of detail to an extreme and making the system design too complex for the management process to get benefit from it. The answer is, unfortunately, it depends!

HOW IS THE FORECASTING ORGANIZATION STRUCTURED IN MOST COMPANIES?

Many organizations concerned about the costs of doing business are looking at forecasting accuracy as a competitive advantage. In these organizations, there is normally a demand manager. The demand manager normally reports to the vice president of sales or the vice president of marketing. This person works closely with the master scheduler and often sits next to him or her. The demand manager would be in direct contact with outside sales people and usually consolidates and massages information submitted by this group. He or she often provides reports to interested parties, such as top management and master scheduling. The demand manager also often feeds the sales and operations planning process with pertinent information.

FINDING OUT MORE INFORMATION ABOUT FORECASTING TECHNIQUES

There are a ton of books available on this topic, but unfortunately most are based on strategies using history and statistics. This can be helpful, but the models become so complex that it gets difficult to build confidence. The real answer lies within the understanding of marketing and selling techniques. There are even more books available on these topics. Understanding the impacts of your company's strategy and actions on customer behavior is the real knowledge opportunity.

There is no easy answer and no magic pill. It takes work and determination. When these two elements exist, there is no limit to the progress. And yes, it is true — *there is no such thing as a perfect forecast!*

This book has free materials available for download from the Web Added Value™ Resource Center at www.jrosspub.com.

13

SOFTWARE TOOLS

The master scheduling tools are almost as important as the master production scheduling process — not quite, but almost. The tools used for master scheduling are normally found within the enterprise resource planning (ERP) business system and are improving every day. These tools make activities like available to promise and order configuration much easier and more accurate. While the master production schedule can be done on a spreadsheet, having it embedded in the ERP business system just makes good sense because of the importance of system integration. In achieving good schedules and performance in customer service, process linkage is important.

ORDER CONFIGURATOR

As is apparent from Figure 13.1, the process linkage is complex. Without software tools to keep the linkage repeatable, it becomes very cumbersome to keep everything in sync. Inside the dotted-line box in the middle of the illustration is the order configurator that links customer information to the master schedule. This must be closely linked to the inventory strategy. The order configurator is much like a hierarchy decision tree that translates customer needs into part numbers with options and features to match each need. Many businesses use inventory strategies other than make to stock. For example, assemble-to-order strategy requires the ability to take options and features and develop a bill of material from the customer inputs that results in a specific configuration that can be understood by the manufacturing operation. As soon as the configuration is understood, it can be translated into a work order and released to the shop for completion. The same would be necessary for make to order except that the manufacturing requirements are a bit more complex.

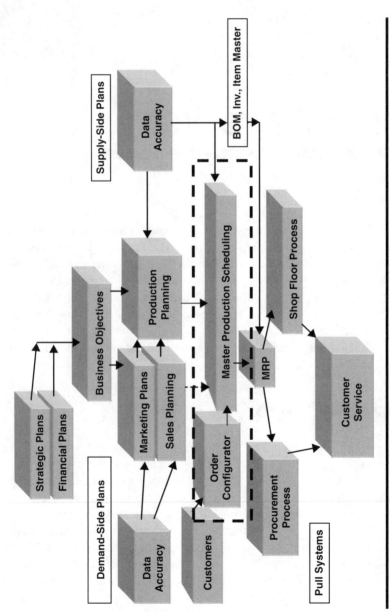

Figure 13.1. System Linkage within the ERP Business System

For the purpose of this discussion, options are those configuration choices that are truly discretionary. In an automobile, for example, a trunk light might be an option. You can buy a car without one, but it is available. Another term that is used similarly is feature. A feature is a choice item that is required. There is more than one possibility, however. A car's engine might be a good example in this category. You can buy an eight-cylinder engine or you can buy a six-cylinder engine, but you cannot buy a car without an engine. A feature is different than an option. Some features are compatible with some options and not with others. For example, a specific light package might be offered with a new car but is not available with the convertible roof feature. (You can buy a hard top or a soft top, but you cannot buy a car without a roof.) It is this sophistication that requires a lot of work up front when defining the configurator. Additionally, some customers may be able to order some configurations and others may not. This is often the case when a manufacturing organization manufactures under various brand names or manufactures products with exclusivity to specific retailers. Every business is a little different in this regard.

The order configurator is a complex element of the ERP business system. It is typical for this piece of the software to have some characteristics specific to the company it is being installed in. It is this author's experience that even in companies with the stamina and fortitude to keep software generic as shipped, the configurator usually requires a few code tweaks. It is very difficult for software suppliers to anticipate every company's order requirements and get them to a satisfactory level in a generic package. The configurator is the translation of options and features ordered by specific customers to a part or item number that can be manufactured and is understandable to the manufacturing organization.

In the example in Figure 13.2, the customer chooses a specific product family and wants it delivered with features F2 and F4 with options Op 1 and Op 2. The order configurator needs to be intelligent enough in this example to know whether this is a realistic configuration. Some options are not available with some features, and again, the configurator needs to be intelligent enough to recognize this. Some systems simply create an edit report for any configurations that are not compatible. Sales personnel then follow up to better understand the customer need. In other environments, customers are simply not allowed to order configurations that are not approved. In such cases, the orders are rejected when submitted. Markets vary as to what the proper protocol might be.

The master scheduler needs to anticipate these various configurations. Sometimes customer or market lead time requirements are shorter than the lead time required to get the product to the customer, in which case planning bills are required to drive options and features to forecasted levels (see Chapter 5 for

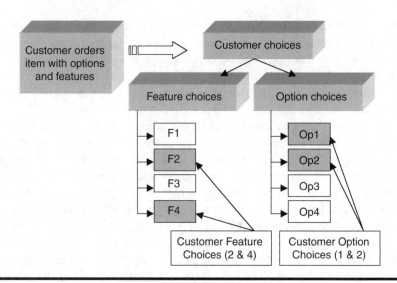

Figure 13.2. Order Configurator at Work

more on planning bills of material). The order configurator needs to be able to link the customer order to the planning process. Sometimes a problem arises in the requirements when a customer order comes in that has a different part number than the planning bill of material. For example, a PC manufacturer may need to plan components for assembly and does so under planning bills of material. When the actual order comes in from the customer, it is for a specific configuration with specific options and features. The software tools have to aid in the consumption of forecasted requirements of the options and features required. Otherwise, the requirements would never be relieved in the system without the master scheduler doing manual maintenance. This happens in many businesses and is a good topic for discussion with a software supplier when considering a new business system.

FINITE SCHEDULING

Just a few short years ago, finite scheduling was usually only done at companies with poor processes. The methodology was used to offset process variation and not usually with complete success. Software has improved in recent years and now is capable of dealing with more complex capacity environments and matching requirements to plans. Like all computer applications, the system tools are only as good as the data fed into the model. If the capacity is overstated, the scheduling tool will overload the manufacturing schedule, and the result will be missed schedules or added costs. There are still huge opportunities to misuse this tool and many still do.

Finite scheduling is the process of filling specific time frames with adequate workloads. The system helps smooth requirements by not allowing the input of more than the capacity can respond to appropriately. Parameters can be set to react to whatever capacity expectations make sense. More sophisticated business systems also allow forward scheduling, which takes into consideration not only the load requirements but also the lead time and sublevel capacity requirements. This is helpful for mix-modeling-type scheduling applications.

Many aspects of finite scheduling get way too much credit, if the truth be known. *Some managers get too excited about software features and forget about the basics of sound business practice. Although helpful, options such as finite scheduling do not take the place of good process.* Available to promise is a forerunner of finite scheduling. Using both finite scheduling and available to promise is a good way to ensure that process controls that provide the right disciplines are in place. Without the disciplines, it is simply a good idea not executed. Another energized topic in ERP business system software today is the advanced planning and scheduling software modules.

ADVANCED PLANNING AND SCHEDULING SOFTWARE MODULES

Many newer ERP business system packages offer what are called advanced planning and scheduling modules. These generally are material requirements planning tools with some new features. These packages all use the basic material requirements planning engine that uses bills of material and inventory records to calculate the requirements to meet forecasted or firm orders. Some also neatly integrate routing records into the bills of material, which allows greater sophistication in internal supply chain management and scheduling.

Advanced planning and scheduling tools often do not actually provide functionality for the whole planning and scheduling process. For example, many systems do not handle the top management planning process known as the sales and operations planning process. In some of these systems, the forecast is loaded into the requirements as gospel, with no consideration of load planning, inventory strategy, or risk assessment. With the popularity of top management sales and operations planning processes growing, more and more software providers are starting to understand this impact and are designing modules to take care of this important need. In the meantime, most of the manufacturers I have worked with over the last 10 years are using Excel or Lotus spreadsheets to fill this gap. With ERP connectivity, these spreadsheet tools and files can provide needed flexibility. Most of the better ERP systems allow ease of spreadsheet use. When integrated, these tools are helpful.

One thing that is helpful about many of the advanced planning and scheduling systems is the ability to manage constraints in the process. As has been said so many times, these benefits are void if the disciplines are not in place to realize the spirit of the software design.

DROP AND DRAG

One of the latest features to enter the scheduling tool world, shortly after GUI (graphic user interface) several years ago, is the ability to drop and drag. This means that as orders are moved in the schedule, they do not have to be rekeyed to move the requirements within the system. Instead, they can be picked up with the mouse and dragged to the slot or desired sequence. The system automatically updates the requirements to the new date and shifts the orders as required. This saves a lot of time for the scheduler and can be very helpful. But making schedule changes too easy can have its downside as well. Changes inside an agreed-to time fence need to follow a handshake governing the "rules of engagement."

The idea of having the full advantage of drop and drag is to have visible linear schedules for some period, even if it is a short timeline. This visibility works especially effectively in make-to-order environments but can also be extremely helpful in make-to-stock strategies. One business that manufactures plastic components uses the drop-and-drag feature in its ERP business system to move product from not only one product line to another but also to another plant when capacity is at risk. Drop and drag in that environment makes the scheduler's job much less complex and tedious. This is a great software tool feature to look for.

ENTERPRISE RESOURCE PLANNING SOFTWARE

All of the system features discussed in this chapter are components of ERP business systems. These business systems are exactly as described — systems tools that are designed to help manage and control planning and resources of an entire enterprise (business). This control, in the more complete business systems, starts with top management planning and ends with shop floor schedule execution controls and supply chain management tools (see Figure 13.3).

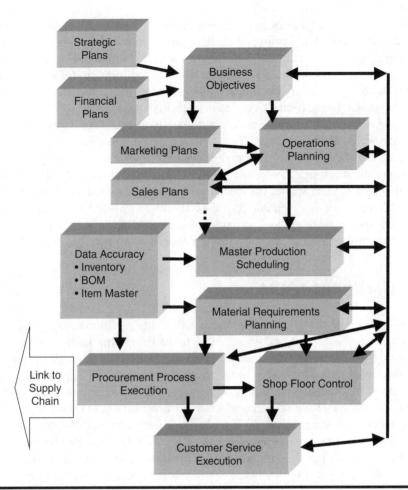

Figure 13.3.   Class A ERP Process Flow Chart

ERP business software systems help with not only the planning process of unit manufacturing but also the management of the financial aspects of the impact of this activity. The business model in Figure 13.3 is a fairly traditional Class A ERP flow. Add to this the infrastructure of financial information hitting the general ledger and balance sheets, and you get close to the duties of good ERP software.

Additionally, many multiplant environments are seeking the ability to drill down into the data cross-organizationally to seek opportunities for efficiencies and best practices. ERP business systems from most of the larger vendors offer this capability today, although it is not as easily done as it sounds and looks in the advertising! In reality, often it is just a common file structure that is allowing reports to interface several data sources within the system and be designed and fed into the data streams. User friendliness is still the challenge with anything other than the simple data query tools that come standard on most ERP systems. I have been introduced to some very popular "easily" programmed (according to the advertisements) data-mining tools that were impossible to use if only occasionally exercised. They are just too complex for normal practitioners to remain proficient at programming new reports without frequent use. I look for the day when these tools are truly intuitive and menu driven, allowing total access to data within the system. It has not happened as of this writing. Nonetheless, existing technology is a giant leap from just a few years ago and a good system planning tool to have in your bag. Data-mining tools are a big asset to an organization bent on driving actions from root cause and facts.

DATA-MINING TOOLS

Possibly one of the most helpful aspects in today's information system management is the (often) add-on modules to ERP systems known as the data-mining tools. These report-generating software tools allow mixing and matching of data from multiple sources to be organized into reports easily retrieved and refreshed by managers with inquisitive eyes. Many of the more popular ERP business systems have not yet perfected this capability at the same level as the specialized software providers. It probably is only a matter of time, however.

Access to data is necessary for data to be valuable. This may be too logical to be profound, but the number one information technology effort associated with ERP systems is normally the development of reports specific to particular departments or teams. With continuous improvement a focus in all high-performance organizations, data access is essential. It is interesting that many of the best ERP software business systems do not have hundreds of reports written and available. Instead, the better ones have user-friendly data query capability, allowing users to make their own reports. I am always leery of software systems that brag about hundreds of reports available. I much prefer easy report writer capability.

Additionally, sophisticated data-mining tools such as Cognos® can be very helpful, especially if a few people in the organization are trained to build the data cubes and design reports. Tools such as these are especially helpful because of the multiple file accessibility that can be built to show data relationships and performance visibility. Many high-performance organizations either own or are moving toward utilizing high-powered add-on data-mining tools. Data are a very valuable asset, especially when leveraged properly.

DATA FIELDS — USE THEM AS THEY WERE DESIGNED!

Many organizations make the mistake of using fields in their ERP systems for other than what they were designed. This would include using the description field for a second-level part number (bad idea anyway) or using a purchasing source code field for defining the stocking strategy for a part or maybe a group code strategy. These types of decisions often seem benign at the time but almost always come back to haunt the team. When updates to the system are offered, these conversions always have to be tracked into the new revision for process compliance and also run the risk of logic changes in the hard code with assumptions made from the original intent of the field. *Just don't do it! Do not use fields in ways they were not designed for by the software producer. This cannot be emphasized enough!*

MANUFACTURING EXECUTION SYSTEMS

The story on software tools and scheduling would not be complete without talking about manufacturing execution systems (MES). At this point, it may sound like a broken record, but, again, an MES offers another opportunity to be fooled into thinking that software glitz can take the place of good process. The MES is one of the more misunderstood products in the software application market. An MES is a scheduling system that makes it easy to drop orders into the schedule and reprioritize existing workloads. This is a good thing when rules of engagement are enforced behind the scenes to disallow costly abuse.

An MES is not typically designed to master schedule and plan the requirements for the 12-month rolling schedule, but it can offer a scheduling advantage. One great application for an MES is in the inventory strategy of assemble to order, in which normally the master production schedule is loaded with subassemblies and demand planning is done on available features and options. An MES is sometimes used to develop the final assembly schedule hourly or daily as customer orders are received. Because of the sophistication of capacity planning functions within an MES, these systems can add to the ease of daily execution scheduling; hence the name. Using an MES to replace netting or master scheduling is often a mistake. High-performance organizations seldom find these systems necessary once robust disciplines are in place. There are lots of examples where an MES has been played up as a huge advantage prior to implementation, only to bring disappointment after fully implemented.

PREREQUISITES OF GOOD SOFTWARE UTILIZATION

Software is a necessary aid in business today. To reap the full benefit of the product data driving the decisions in the software, data and application of the information need to be reasonable and accurate. This is the well-worn "garbage in, garbage out" saying. It has never been truer. As manufacturing companies lean their processes and reduce the number of eyes watching them, reliance on strong and accurate data sources becomes critical. It cannot be said often enough that software tools are not the panacea for bad execution of process or poor discipline.

Areas of specific interest would include the following:

- Bill of material
- Inventory location balances
- Lead times
- Cost standards
- Routings
- Demonstrated capacity

The way to get at these requirements is through good process design, management systems, accountability, and robust measurements. Leadership must believe in this resource for it to be a reality. *Software offers no substitute for good process, but good process can be enhanced by good software tools.* Once process is predictable and robust, good tools can increase the efficiency or act as the enabler for efficient process.

This book has free materials available for download from the
Web Added Value™ Resource Center at www.jrosspub.com.

14

MASTER SCHEDULER JOB DESCRIPTION

By this point in the book, it should be evident that the master scheduler's job is pretty important in most manufacturing businesses. Getting the right person for the job is imperative. Some of the most critical aspects to look for in a person as he or she starts the job of master scheduler are as follows:

- **Attention to detail** — A master scheduler must be able to detect problems before the issues arise. Having an appetite for detail is a real asset.
- **Leadership skills** — Leadership is a trait that allows the master scheduler to lead an entire facility or business through a schedule. Leadership skills are essential.
- **Knowledge of the products and processes** — It is difficult to hire a master scheduler from the outside and have the person be effective quickly. The best master schedulers have good product knowledge and understand the idiosyncrasies of the business and markets.
- **Respected by manufacturing and sales** — This trait can be difficult to come by, but it is worth a lot when you can achieve this harmony. People who have been in the business for a while and have worked in a few positions make good candidates. This respect must also come from top management.
- **Good communicator** — Because the master scheduler is the liaison between the demand and supply sides of the business, it is important that he or she can effectively communicate both in writing and verbally.
- **System expertise** — Good business systems expertise is a must. This includes being able to build planning bills, maintain schedules effi-

ciently, and link the sales and operations planning (S&OP) to the schedule, frequently updating it. Enterprise resource planning systems are functionally similar from one brand to another, but the specifics are often very different in how these functions are accomplished from screen to screen. Having systems knowledge in the appropriate system is worth quite a lot.

- **APICS certification is always a plus** — The experience and know-how required to pass the tests do not paint the entire picture, but are good indicators.
- **Honestly and integrity and a good attitude toward the job and the business** — This may go without saying, but with the responsibility given to the master scheduler, it is important to be able to trust and depend on the master scheduler to make the right decisions and to have the judgment to bring top management in on decisions when appropriate.
- **Problem-solving skills** — A good master scheduler will have good problem-solving skills and be well versed in tools such as brainstorming, 5-why diagramming, Pareto analysis, fishbone diagrams, and process mapping.

BACKGROUND

Master schedulers come from all backgrounds. The following list indicates backgrounds of people I have worked with who made great master schedulers:

- Materials manager
- Demand manager
- Production manager
- Maintenance manager
- Sales manager
- Industrial engineer
- Product planner

As you can see, the backgrounds of this group are so varied that background alone may not be that important. What is important is the ability to command respect and lead people. Product knowledge is also important, as is knowledge of the process idiosyncrasies of the firm. Also notice that the prerequisites for this job are generally of a management nature. This type of background helps as master scheduling is truly a key management position. Many master schedulers have other functions reporting to them. Figure 14.1 shows a typical yet simple

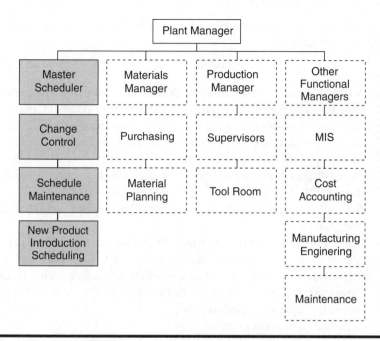

Figure 14.1.   Typical Manufacturing Plant Master Scheduler Reporting Structure

master scheduling organizational reporting structure within a multiplant multi-million-dollar operation/environment.

MASTER SCHEDULING JOB DESCRIPTION

The job description will change depending on the environment, organization, and market complexity, but the following outline can be helpful in getting started on documenting the master scheduler's duties.

Title: Master Production Scheduler   **Date Written**: 1/15/200X
                                              Written by: Jane Small

Reporting Structure: Reports directly to the plant manager. Generally has up to 15 reports consisting of planners, administrative people, and schedulers.

Interaction with Other Areas within the Business: The master scheduler must interact with all levels of management within the organization at a very detailed level. Interaction can also happen on a periodic basis with customers and other outside representatives.

Duties
1. Maintain master production schedule, including no past due requirements
2. Develop S&OP reports regarding demand and operations plans
3. Report performance on plan accuracy to management
4. Determine tie-breaker calls during schedule conflicts
5. Maintain planning bills of material
6. Maintain schedule linkage between the master production schedule and the S&OP plan
7. Schedule all requirements into the schedule (service, actual, distribution, forecasted, etc.)
8. Communicate with sales regularly to manage order sequencing
9. Manage costs by understanding changeover effects and planning accordingly, balancing inventory costs with (lack of) manufacturing flexibility
10. Develop and maintain the rules of engagement for order management, including master scheduling policy
11. Schedule new product pilots
12. Oversee scheduling of upcoming new product introduction
13. Assess risks within the schedule and communicate appropriately
14. Maintain accuracy in the scheduling process
15. Review and propose all inventory strategies (MTS, ATO, MTO, ETO, etc.)

Metrics Reflecting Performance
1. Delivery to request — Percent of complete orders that deliver on time to the original customer request
2. Delivery on time to the customer — Percentage of complete orders that deliver on time to the customers to the original promise date
3. Ship on time to the customer — Percentage of complete orders that ship on time to the customer to the original promise date
4. Daily schedule adherence — Percent of full quantity orders that are completed on the day they were scheduled
5. Weekly schedule adherence ——Percentage of full quantity orders that are completed within the week they were scheduled
6. Schedule stability — Percent of scheduled orders that do not have dates changed within the fixed fence
7. Inventory turns — Number of times per 12-month rolling period that inventory is turned over based on existing inventory dollars and forecasted dollar requirements

The job description in your business may differ from this sample, but many of the concepts should be included. Most of the ideas listed would be appropriate in almost any master scheduler position within manufacturing. By attending local APICS dinner meetings, you should be able to meet others who have master scheduler job descriptions. Swapping ideas can keep the quality up and the workload down. This lesson is true in many related topics.

This book has free materials available for download from the
Web Added Value™ Resource Center at www.jrosspub.com.

15

CLASS A ERP BUSINESS MODEL AND PERFORMANCE

Class A ERP is probably not as well recognized as a performance model as it should be. Enterprise resource planning is a business model that is all-encompassing within a manufacturing business. The differentiator for making the ERP model Class A is *performance within* that business model. Class A ERP is the performance standard that defines high-performance discipline, accuracy, and customer service in the ERP process linkage. These criteria are defined by metric definitions and performance levels within those measures, roles of process owners, project management, accountability infrastructure, and customer service levels. As we get deeper into this topic, the details will be more specifically documented.

MATERIAL REQUIREMENTS PLANNING

Class A ERP did not start out in full bloom. Early thinkers in this space would certainly include people like Oliver Wight and George Plossl. I first met George at The Raymond Corporation in the 1970s. The Raymond Corporation is a major player in the manufacturing of narrow-aisle material handling equipment and also the company where I "grew up." George had been a previous associate of Jim Harty, who was the CEO of Raymond at the time. All three of these gentlemen worked together years before at the Stanley Works. Oli and George, who both leveraged their thought leadership into consulting businesses, are usually included in descriptions of the "fathers" of inventory and production control. Mr. Harty continued to stay in the practitioner side of business.

Material requirements planning (MRP) was the first formal recognized process that started to link the manufacturing planning processes. At the time, it was leading edge. MRP helped inventory management by time phasing requirements and allowing the utilization of computer processing of information as it related to forecasted and firm orders. Weighted formulas for calculating requirements could then be utilized rather than committing entire bills of material for each order to shorten lead time.

MANUFACTURING RESOURCE PLANNING

Early to Mid-1980s

Manufacturing resource planning (MRP II) happened as many evolutions do — due to need. We, as humans, do not change radically unless we are pushed in some manner. In the 1980s, manufacturing was getting pushed in a big way. Interest rates were skyrocketing and businesses were under pressure from many fronts. Competition from other parts of the world was becoming more real, and prices reflected it. The need to cut costs and therefore become more competitive drove materials management professionals to look for better and more efficient methods to plan material deliveries. It was a matter of survival. This effort led to more discipline on data inputs and files accuracy.

In the 1980s, because of cost constraints and competitive pressure, more emphasis was developing in businesses in the areas of capacity planning and increased scheduling disciplines. It was becoming more and more important to build and stock the right inventory. By the mid-1980s, the master production schedule (MPS) had been born.

The MPS was a schedule just like so many others that had existed over the years, but with a new management approach. Capacity was recognized as a defining constraint, and master scheduling received new support as businesses started to recognize reality more frequently. Although evolution of production control happened over several years of need, it was about this time when the American Production and Inventory Control Society (APICS) and people associated with the field of manufacturing planning started to refer to the business planning model as MRP II. With the new process focus came various defined levels of proficiency. By the mid-1980s, a few consultants had developed from the field. Each had a view of what proficient MRP II performance was, but for more than one reason, similarities developed. APICS probably played a big part in the cross-pollination, as most of the "early experts" became regulars on the speaking circuit. Interestingly enough, all of the materials masters and many of the best speakers in the world on topics of interest to the reader of this book got their start in this APICS road show circuit. This speaking circuit includes APICS dinner meetings, regional seminars, and the crème de la crème, the Annual APICS International Conference and Convention.

From this group of traveling missionaries on the topics of planning and control, a few became household names in the APICS circles. Some of the ones I remember most from that era are Oli Wight, George Plossl, David Buker, Jack Gips, Eugene Baker, and Dave Garwood. While many would credit Oli Wight for the early definitions of Class A, in reality it was probably the sharing of ideas and experiences by this entire crowd that birthed the new performance definition. By the late 1980s, the standard had become fairly consistent, consultant to consultant. I emphasize *fairly consistent* because there was and is no true worldwide accepted Class A MRP II or Class A ERP standard.

MRP II certification in those early days was simple. It required measures to be audited and confirmed at certain levels. The early measurements were variations on the following reasonably consistent list:

Measurement	*Minimum Performance Requirement*
Profit accuracy	90 then later increased to 95%
Sales forecast accuracy	85 then later increased to 90%
Production plan accuracy	90 then later increased to 95%
Schedule accuracy	90 then later increased to 95%
Inventory record accuracy	90 then later increased to 95%
Bill of material accuracy	95 then later increased to 98%
Routing record accuracy	90 then later increased to 95%
Shop floor control accuracy	90 then later increased to 95%
Supplier promise accuracy	90 then later increased to 95%
Customer promise accuracy	90 then later increased to 95%
Overall performance	90 then later increased to 95%*

Mid to Late 1980s

The top management planning process was developed earlier, but it was during this period that the process coined as sales and operations planning (S&OP) by

---

* Sustainability normally was required and proven by showing 90 to 95% performance overall for at least three months.

the Oliver Wight group became a high-level honed top management planning process. Top decision makers in the business could now more directly influence activity at the plant level, including inventory levels and general risk management involving capacity and anticipated demand.

The measurements were fairly loosely defined and could differ widely from business to business and one consultant to another. Nonetheless, Class A MRP II was quite popular as word got around about the benefits companies were seeing as a result. The early to mid-1980s period was the heyday for MRP II consultants. David W. Buker, Inc., of Antioch, Illinois, was named during this period as one of the fastest growing companies in America by a leading business magazine. Oli Wight first, and later David Buker, developed video education in this space that helped deal with the growing need for education and training of the masses within the manufacturing world. These videos sold for thousands of dollars based on the return that many businesses enjoyed. Additionally, public classes on the topic delivered by the masters were filled regularly, and businesses in need of competitive advantage set out to implement this new formula for success. Needless to say, some companies were much more successful than others. In most cases, there were gains, even with this rudimentary approach, because the general population of manufacturing businesses at the time had extremely poor discipline in scheduling and data accuracy; even half-hearted attempts at Class A MRP II implementation yielded benefits. The benefits commonly included:

- Reduced inventory
- Increased customer service
- Higher productivity
- Lower costs

Early to Mid-1990s

In the 1990s, especially the mid-1990s, the Class A model evolved from a set of metrics to a process with several required components. Various successful implementations as well as other evolving methodologies, including ISO (International Organization for Standardization), total quality, the Malcolm Baldrige National Quality Award, and even just in time, all influenced the Class A MRP II process.

ISO, which in the early days had no more value than to fill a political gap in trade agreements among the newly formed European Union countries, even contributed in some small way as documentation of process gained more respect in securing repeatability through process control. During this time, thought leadership within Class A MRP II teaching developed both a management infrastructure for accountability and additional business model structure. The first Class A performance objectives centered around objectives in three areas of focus — (1) process design, (2) management system, and (3) results — started to affect behaviors and aid in process control.

During the early to mid-1990s, the implementation times of Class A performance and structure started decreasing. Most organizations serious about their implementations were able to successfully reach Class A performance levels within 12 to 18 months. The efficiencies came from top management being more involved and follow-up being more predictable and planned. When process owners take the time to do root cause analysis and drive actions, this kind of improvement can happen faster! Most organizations were also finding that education, the investment in human capital, was often worth the monetary investment when topics were carefully chosen. This was especially relevant in organizations that were upgrading their software tools.

Software companies had a growth period in the 1990s as computer technology and process evolved quickly. It was during this period that the term enterprise resource planning and the acronym ERP were frequently marketed to differentiate one software package from another. ERP became popular, and by the late 1990s, due largely to software marketing strategies, the term MRP II became yesterday's newspaper. The process requirements for Class A performance continued to evolve.

Late 1990s

Because the scope of Class A performance was growing with the competitive pressure of most manufacturing markets and admittedly because the market was asking for ERP rather than MRP II, in the late 1990s Class A upgraded its cloak to Class A ERP. As the competitive pressures increased from global markets, so did the need for process improvements in manufacturing. Class A ERP in the late 1990s was becoming quite refined, especially as compared to the early days of MRP. By now, the Class A process had elements of expectation not only in scheduling and planning but also in inventory strategy, quality, demand forecasting, supply chain management, project management, and plan execution. The Class A ERP management system had grown into monthly, weekly, and daily elements and expectations. Not only were monthly top management planning events scheduled well in advance, but there were weekly and daily regimens required by high-performance organizations. On the other side of this matrix, a predictable accountability infrastructure was scheduled in 6-month intervals as well as 12-month schedules, all dictated by the evolving Class A ERP process criteria. These management system events included the following:

Yearly
- **Strategic review** — Updates to the strategic plans
- **Business imperatives** — Prioritizing the short list of "must-do" objectives
- **Talent review** — Management assessment of key employee skills
- **Succession planning** — Key position and skills analysis of bench depth

Monthly
- **S&OP** — Risk management of capacity, inventory, and customer service decisions
- **Project review** — Reviewing progress on business imperatives

Weekly
- **Performance review** — Process owner review of progress
- **Project review** — Detailed review of projects by process owners
- **Clear-to-build** — Handshake between the MPS and production managers

Daily
- **Schedule review** — Detail review of yesterday and today's requirements
- **Daily walk-through** — Management-by-observation tour of the factory

In each case, these management system events helped to control the speed and sustainability of conversion to Class A ERP performance. Class A ERP was becoming much more predictable, company to company and consultant to consultant, even though there was no world standard such as ISO or Six Sigma. Implementations in some instances were now, for the first time, taking no more than six months from initial Class A ERP education to certification.

In the mid-1990s, AlliedSignal (now Honeywell) was in the process of converting to SAP business management software. It was a vision of those in management at AlliedSignal to have the disciplines in place prior to the implementation of new software. They had the insight to use Class A as the standard. At the time, there was enough standardization accepted within Class A that three separate consulting companies were approved as sources for this training and certification. The management vision was communicated to the plants; it stated simply that the new software would not be available except at facilities that had met Class A criteria to a certain level. This proved to be a good prerequisite for software performance. This effort resulted in significant improvements in inventory control and accuracy, schedule disciplines and execution, and ultimately customer service.

Topics like transaction design were addressed and any process bugs worked out prior to turning on the new software. Data were measured, scrubbed, and maintained through Class A management systems. It took those process variation areas that are common in many organizations off the list of software implementation issues.

The new acronym ERP was emerging as the replacement for MRP II. Lots of consulting groups and software companies would like to take credit for the innovation, but my recollection is that the software companies, again, were the group that marketed this migration, and the term stuck due to their efforts and marketing noise level. In any event, it made a lot of sense. The message was that supply chain management was much bigger than MRP II. Use of the word "enterprise" suggested a much bigger planning scope. It was time a change. ERP had the right mix of new ideas and proven processes to be the successor to MRP II.

THE CLASS A ERP STANDARD OF TODAY

Because this book focuses on one of the most important elements of Class A performance, master scheduling, it is important to describe at least the main principles of Class A ERP. Boiled down, Class A ERP is about proficiency in all of the following areas of focus:

1. Prioritization and management of business objectives
   A. Project management
      i.   Project funnel
      ii. Prioritization of projects
      iii. Resources and skills required
      iv. Review process
   B. Human capital management and investment
      i.   Professional society affiliations
      ii. In-house education
      iii. Tuition aid programs and guidelines
      iv. Training
         a. New employee training
         b. Exiting workforce training
         c. Skills assessment
   C. Business imperatives
      i.   Hoshin planning
      ii. Review process and documentation

D. New product introduction
E. Accountability infrastructure
   i. Metrics
   ii. Management systems
      a. Daily
      b. Weekly
      c. Monthly
2. S&OP processes
  A. Strategic planning
     i. Markets
     ii. Core competence
  B. Demand planning
     i. Mix
     ii. Volume
  C. Operations planning
     i. Supply chain partnerships
     ii. Capacity planning
       a. Internal capacity
       b. External capacity
  D. Financial planning
     i. Profit
     ii. Capital spending
     iii. Revenue
3. Scheduling disciplines and production planning
  A. Master scheduling
  B. Rules of engagement
4. Data integrity
  A. Inventory location balance accuracy
     i. Warehouse design
     ii. Transaction design
     iii. Point-of-use storage
     iv. Location design
       a. Raw
       b. Components
       c. Work in process
       d. Finished goods
     v. Cycle count process
       a. ABC stratification
       b. Tolerances allowed
  B. Bills of material or bills of resource accuracy
     i. Engineering change

      ii. Process to repair bills of material
      iii. Audit process
      iv. Routing linkage to bill of material
    C. Item master accuracy
      i. Lead times
      ii. Cost standards
    D. System security
    E. Part number design
5. Execution of schedules and plans
    A. Procurement process
      i. Linkage to MPS
      ii. Supply chain communications process
      iii. Management systems
    B. Shop floor control
      i. Linkage to MPS
      ii. Communications process
      iii. Management systems

In the 2000s, the metrics would be as follows:

Measurement	Minimum Performance Requirement
Profit and/or budget accuracy	95%
Sales forecast accuracy by product family	90%
Production plan accuracy by product family	95%
MPS accuracy	95%
Schedule stability	95%
First-time quality	97%
Inventory record accuracy	98%
Bill of material accuracy	99%
Item master accuracy	95%
Daily schedule adherence	95%
Procurement process accuracy	95%
Customer promise accuracy	95%
Overall performance	95%*

Not only does the number of metrics continue to increase, but so does the threshold of acceptability. These metrics are built from an ERP business model that shows succinct linkage between levels and activities, the essence of Class

* Sustainability normally was required and proven by showing 90 to 95% performance overall for at least three months.

A ERP. Class A ERP is not about the metrics. It is much more than that. It is about the process design, the management systems to ensure the sustainability and improvement of the process execution, and lastly the results or performance. The results are evidenced by the metrics.

In Figure 15.1, you can see the whole manufacturing organization, from top management planning to shop floor execution. The arrow from the procurement process indicates where the link would exist between this model and the same model in the supplier's business. Keep in mind that this model is applicable in any business, from process flow businesses to sheet metal shops. Class A ERP will work in any business, including your suppliers'.

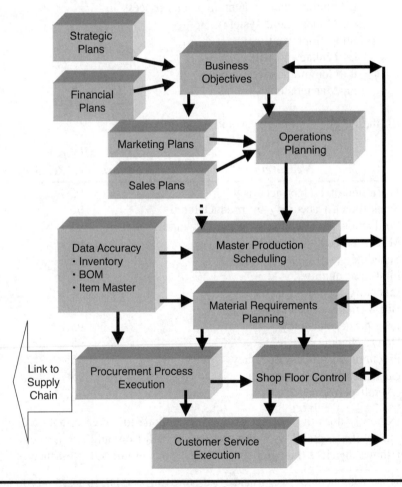

Figure 15.1. ERP Business System Model

Having the ERP business models linked is the essence of supply chain management. Acknowledging the rules, roles, accountability, and information flow is a major element of Class A ERP performance. The gains come from disciplines of process as well as shared goals.

CLASS A ERP CERTIFICATION

Certification is a confirmation from an outside party that you have met the rigid criteria of Class A ERP performance. The criteria are areas that correspond with the criteria listed earlier in this chapter. There is a certification audit document that is several pages long which is administered by an experienced auditor. Companies will often prefer to have an audit done a month or so prior to the final one to get a "gap analysis" punch list. Since Class A is so all-encompassing, this often makes a lot of sense. At DHSheldon & Associates, we generally follow a consistent agenda.

Class A ERP Audit Agenda

1. Walk-around
   A. Review 5S housekeeping and workplace organization
   B. Observe "visible factory" communication boards
   C. Discuss daily performance with a worker randomly selected from the floor
2. Review of Class A–related documentation
   A. Check for key points important to the sustainability of Class A
   B. Check for good process design, especially in areas of master scheduling, S&OP, inventory accuracy, management systems, and measurement processes
3. Observe the top management S&OP process
4. Discuss and observe evidence of the strategic planning process
   A. Review documents
   B. Observe the business imperative list
5. Spend time reviewing each process with the process owner for the following processes
   A. Master scheduling
      i. Rules of engagement
         a. Practices and communication
         b. Time fences
      ii. Disciplines
      iii. Measurement audit

B. Materials planning
   i. Rules of engagement
   ii. Measurement audit

C. Purchasing
   i. Check for past due purchase orders
   ii. Check for weekly maintenance of open purchase orders
   iii. Review metric
   iv. Review supply chain relationships and partnership agreements
   v. Review reverse auction process

D. Shop floor schedule attainment
   i. Review metric
   ii. Review process for assigning daily schedule
   iii. Check for linkage to the MPS

E. Education and training
   i. Review policy and practices
   ii. Review documentation of employee education and training
   iii. Review skills assessment process
   iv. Discuss involvement with professional societies and review policy regarding same
   v. Review policies for tuition reimbursement and educational support for off-campus education
   vi. Review new employee training policies and execution of same

F. Project management process
   i. Review project funnel process
   ii. Review prioritization process
   iii. Audit top manager (within the facility being audited) project review process
   iv. Understand project linkage to business imperatives of the business

G. Quality
   i. Review first-time quality metric

H. Bill of material accuracy
   i. Review metric
     a. Specification
     b. ERP business system record
     c. What workers do on the factory floor
   ii. Review engineering change practices
   iii. Audit average cycle time for making changes to inaccurate bills of material discovered by the business

6. Customer promise accuracy
  A. Review promise process
  B. Review metric

7. Management system — accountability infrastructure
   A. Review documentation
   B. Observe weekly performance review
   C. Observe daily walk-through or daily schedule adherence meeting
   D. Observe weekly clear-to-build process
8. New product introduction process
   A. Review process gates
   B. See evidence of disciplines
   C. Review past postlaunch audits done internally for past product introductions
9. Inventory accuracy review
   A. Review transaction processes
   B. Review cycle count process methodology
   C. Audit raw and components storage accuracy
      i. Take sample counts and check accuracy to the system's perpetual record
      ii. Review cycle count sheets from recent factory counts for process integrity
   D. Audit finished goods accuracy
      i. Take sample counts and check accuracy to the system's perpetual record
      ii. Review cycle count sheets from recent factory counts for process integrity
   E. Review work-in-process accuracy management system
      i. Check definitions and applications of "controlled inventory areas"
      ii. Check for the proper application of the "rule of 24." (The "rule of 24" has to do with the length of time in hours that an SKU is planned or by practice kept in one area at a time. If it is more than 24 hours, the perpetual record in the ERP system should be updated as a regular practice at the time of the physical inventory transfer.)

The last items for discussion within certification are the focus areas around each topic. During the certification, the auditor is looking for three different aspects of each topic:

1. Robust process design
2. Management systems to make sure it is and will continue to happen
3. Results that are demonstrated by the metrics

In addition to these three areas of focus, there is the concern of sustainability. Organizations must prove that they can sustain Class A ERP levels of performance. This is usually proven by three months of sustained results or clear trends showing consistent improvement in the results over time.

Class A ERP Certification: Robust Process Design

In the process design review, the auditor is looking for proper and acceptable procedures evidenced by good documentation. An example of noncompliance to Class A ERP criteria would be if the organization did not have proper transaction documentation for inventory transactions. Another noncompliance example might be if the transaction design included too many transactions or was too complex for easy training and execution.

Class A ERP Certification: Management Systems

The second level of audit is the management systems review. The focus here is to ensure that the processes are set for sustainability. Normally, that means there is some fool proofing, "go, no-go" check, or accountability infrastructure. Continuing with the inventory transaction example, a noncompliance could consist of a well-designed transaction process for inventory transactions without any accuracy check such as a cycle count program or inventory accuracy audit. Most Class A processes are governed by the weekly performance review meeting where all process owners report progress and improvement of their process once a week, often on Tuesday at 1:00 P.M. By having a management system in place, good process can continue through generations of employee turnover.

Class A ERP Certification: Results

The component of Class A ERP that most people think of first is the metrics. Organizations often think they are ready for certification just because the metric performance is at Class A ERP levels of acceptability and are surprised by the shortcomings in process design and management systems. As you can see from the descriptions above, both of those areas of focus are important for high-performance organizations. Nonetheless, performance is where the rubber meets the road. Everybody can relate to measurements. We all have them in our lives, one place or another, and most of us have several. Class A ERP has numerous measurements, and to meet certification requirements, they need to be at Class A minimums and need to show sustainability.

The measurements are only part of the picture, but obviously an important one. The certification audit generally not only looks for the correct performance, but also looks at actions and trends to see if the process is tracking a normal curve and shows ownership and sustainability.

This audit process introduction should give you a sense for the profundity of the Class A ERP focus. At this point, you should also have an appreciation for the benefits of achieving it. While Class A ERP may not be the equivalent of world class performance, it is impossible for any organization to achieve Class A ERP performance without being a high-performance company.

This book has free materials available for download from the Web Added Value™ Resource Center at www.jrosspub.com.

16

INTEGRATING THE MASTER PRODUCTION SCHEDULE WITH LEAN AND SIX SIGMA

Both lean and Six Sigma are topics that get a lot of press these days, and for good reason. These two topics, along with a third, enterprise resource planning (ERP), where the master production schedule (MPS) is prominent, make up a logical progression toward world class performance. Each has a specific approach and focus, and each of these process methodologies, done correctly, increases the success of manufacturing organizations. The master schedule plays an important role in businesses using any or all of these methodologies. While there is no conflict whatsoever, there is still a lot of confusion in the marketplace about how ERP, lean, and Six Sigma fit together. One of the best ways to think of these process focus areas and the sequencing is to picture them in a step chart (see Figure 16.1).

The first focus high-performance companies normally achieve is ERP process capability. ERP process capability is the predictability of process — when a company says it is going to do something, it is done. This would include master schedules, detailed plans, and customer service, but would also include internal practices like data accuracy. Class A ERP performance is a specific level of proficiency in areas that include top management planning, operations

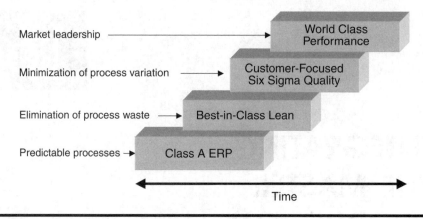

Figure 16.1. Journey to World Class Performance

planning, data accuracy, and execution of plans. The MPS is an integral part of the ERP business system — one of the most important, and therefore many of the basic MPS process requirements are a first-level priority even before flexibility and speed become the leading topic. High-performance ERP is about predictability and repeatability. The master schedule is the heartbeat of this process. Once companies have this level of predictability, the next logical step is to lean the processes to optimize performance of them.

In the lean methodology, the focus is on flexibility and speed. Much of this is gained through the elimination of waste (see Figure 16.2).

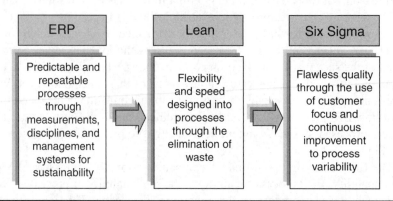

Figure 16.2. Performance Sequence of Normal High-Performance Process Maturity

The whole idea behind separating the process improvement approaches is simply to define the journey into bite-size pieces and celebration points. The segregation points become:

1. **Becoming predictable** — Making promises and keeping them. When meeting market requirements, promises can make a big difference in competitive advantage, but they are not the whole picture. By focusing on process capability first, processes are reinvented, redesigned, or made repeatable through the use of metrics, management systems, and root cause analysis and barrier elimination. The deliverables from this element of process improvement are the management system and the cultural shift to measurements and analysis. The MPS is in the middle of this process focus with performance data being analyzed daily, weekly, and monthly. With the Class A or high-performance ERP emphasis on the sales and operations planning process and supply chain management, there is less need for prioritization of projects. It happens naturally as a result of the management systems. Because the methodology is well defined, the process becomes a very efficient cultural-shifting strategy. It is not, by any means, the whole answer, however.

2. **Improving the processes by eliminating waste** — Lean focus works well for the second step in the journey to excellence as the emphasis naturally shifts from process design and capability to process improvement. Admittedly, there are lots of examples where the processes get redesigned a second time in this space. To think that there is ever an end to process improvement is ludicrous anyway. This second look may not be the last time a specific process gets completely redesigned. The emphasis is different with a lean focus however. The emphasis in this space shifts from only process capability to process flexibility, speed, and responsiveness to customer need, a step above the normal ERP emphasis. The ERP prerequisite deliverables of the daily/weekly/monthly management system and habits of measurement also make lean very efficient as the emphasis moves from predictability to repeatable speed. At the celebration point from lean, companies can expect fast and responsive processes with measurements and management systems that will continue the improvement into the Six Sigma space. This is where improvements such as setup reduction and small lot emphasis are made. Obviously, this impacts the MPS in several ways, as the MPS is the drumbeat of the manufacturing process.

3. **The third progression involves lowering costs further by minimizing process variation** — All processes have variation. This third step is one

that, in the best organizations, never ends: Six Sigma. Process variation in all organizations and processes is unavoidable, but in high-performance organizations, *decreasing* process variation is also unavoidable. In statistical terms, one sigma is equal to one standard deviation (sigma [σ] is the statistical symbol for standard deviation). Recall from college statistics that standard deviation is a statistical measure of variation; therefore, as process variation is decreased, so is the standard deviation of that process. Customer requirements are the limits of acceptability in any process, and as the standard deviation is decreased, the likelihood of missing customer requirements is also decreased as long as the average of the process is located within the upper and lower specification ranges. This is the theory behind the Six Sigma capability measurement. If a process repeatedly results in Six Sigma quality, the likelihood of a defect is almost unheard of — 3.4 defects within a million opportunities.

LEAN IN THE MASTER PRODUCTION SCHEDULE

Lean is both a methodology and a goal for performance. With the focus almost entirely on elimination of waste and increasing flexibility and speed, lean thinking starts from the premise that there is waste all around us. Waste is generally found in eight different areas, all of which can affect the MPS:

1. Unnecessary material movement
2. Unnecessary worker movement
3. Wasted space
4. Wasted time
5. Wasted material
   a. Scrap
   b. Rework
   c. Lost
   d. Damaged
   e. Mistakenly ordered
   f. Unnecessary
6. Wasted effort (poor quality)
   a. Work quality
   b. Material quality
   c. Process design quality
   d. Standards
7. Wasted knowledge or information
8. Wasted creativity

Let's look at these one at a time and apply the logic of lean integration within the MPS process. It really is a very good fit.

Unnecessary material movement — The first topic of waste is found in every corner of almost every business. If we start with the premise that *any* material movement is an opportunity for *elimination* of movement, a lot of opportunity arises. If capacity is at a premium and proper metrics are in place, material movement opportunities will be flushed out every time as the result of metrics driving root cause analysis and resulting actions. This is helpful to the MPS process, allowing the most efficient scheduling to meet even the most unruly customer requests.

Unnecessary worker movement — Again, when the unneeded movement and associated time are eliminated from the manufacturing process, only efficiency can result, making the MPS process more robust.

Wasted space — In many high-performance organizations, growth is a given. Space can be at a premium. Reducing inventory buffers is a logical answer, but unless the processes are in place to allow for the responsiveness, the buffers are necessary. Space occupied by inventory that is only needed because the lot sizes are too big or the setups are too long is waste on many levels.

Wasted time — A major focus of lean, wasted time is one of the first elements of waste identified in the MPS metrics. Rate or uptime is a critical component of the MPS process metric and of schedule adherence in general.

Wasted material — Yield is a common MPS concern. The more predictable the yield, the easier it is to schedule the MPS. Scrap can also be a frequent issue when driving inventory balance accuracy. Often, it is not correctly accounted for and results in inflated balances and inventory shrink. Data accuracy normally drives this to the surface very early in the process. Rework is another situation resulting in material being wasted, along with labor, resulting in schedule inaccuracy.

Wasted effort — Inappropriately expended labor such as in poor quality is quickly identified through a first-time quality metric. Process design quality is one area that gets special attention through the bill of material and standards review requirements in Class A ERP. All of these areas affect MPS accuracy and process.

Wasted creativity — Imagination has always been a favorite waste opportunity of mine. When businesses engage the entire workforce, it becomes enjoyable to see ideas generated. Creativity comes in many forms. Some of the best ideas I have seen implemented were not thought of by management or engineering, but by people from the office or factory floor. Engaging the entire workforce (from the head down — not just their backs) is a big step forward in asset utilization and another clear integration with the resources of the enterprise. Bob Shearer (one of the master schedulers interviewed in the next chapter) uses an organization chart that now includes order management, a group that normally reports to sales. This was the result of people being engaged in the process to figure out better approaches. Sometimes the best answer is not what is expected from the norm. This does not happen without creativity and imagination.

LEAN TOOLS

It is not the intention of this book to be a guide to lean. There are other books that do that. It is the intention of this book, however, to show how lean easily and logically integrates with the MPS. Robust MPS practice always delivers metrics as a main driver of action and analysis in the business. Without good problem-solving tools, the actions would not be a result from the metrics. It is the analysis that leads the way. Lean tools are at the center of this analysis. Process mapping is an unavoidable tool in a high-performance organization. Several of these tools will be detailed later in this chapter, including value organization alignment map, time value map, swim lane flow chart, physical process map, and logical process flow. These various mapping tools include several important methods to identify waste in a process.

If after analysis and improvement a master scheduling process is at the 95% capability level, the expectation is that 95% of the time this process will result in acceptable performance. This does not necessarily mean, however, that the process is without waste. The second-level look at this process may find that there is an opportunity to improve the communication, lock the schedules at some predetermined lead time, include predictable process variation of some operations, etc. This second-level look is often the look through a "lean" filter lens.

Process mapping allows this process to be looked at for its elements. During MPS investigation and improvement activity, it might be determined that there was some discipline missing in the supply chain feed. Other metrics and associated actions would result, and the supplier would become engaged in the improvement process, allowing the improvement of the MPS accuracy. It is all interconnected!

After the process is repeatable to an acceptable performance level, the next level of improvement must take place. That level would most likely also begin with a process map. After scrutiny and asking whether each of these steps is value add, the resulting process could be much different. Without knowing the specifics, this could be the right and appropriate next step in the improvement process of eliminating waste. Clearly, lean is playing an important role in the continuous improvement process for high-performance master scheduling. The job of continuous improvement is never done — never!

SIX SIGMA

The third leg of the improvement journey is often Six Sigma (see Figure 16.2). With only 3.4 defects allowed in a million opportunities, the standards are higher than most objectives in ERP requirements or even most lean measurement standards. Variation-based metrics are superior to percentage performance metrics as they define the swings in variation — how far out of spec the percentage of misses go. When using just percentage measurements, one can only see that some of the process results are out of spec, not how much these results are off. This is part of the concept behind the progressive, step-by-step improvement approach. Class A ERP has a 95% threshold; Six Sigma has a 99.99966% threshold. There is great value in ratcheting the objectives.

Six Sigma is actually a definition of the levels of defects found in a certain process. Six Sigma, as stated earlier, means there are predictably 3.4 defects for every million opportunities in a given process. In reality, Six Sigma has also become a label for a specific methodology of problem solving and project management. According to many Six Sigma experts, the methodology label was started by Motorola and made famous by General Electric. Jack Welch, former CEO of General Electric, was one of the biggest advocates for this methodology and helped make it popular at many companies. Most large companies have some form of Six Sigma process improvement methodology in process. Some of the more successful companies I have interfaced with that utilize Six Sigma somewhere in their strategy are Dell, Lockheed Martin, GE, Honeywell (formerly AlliedSignal), Motorola, NCR Corporation, E-Z-Go, and hundreds of others.

Because all processes have variation, a methodology to minimize it would make a lot of sense to most people. Master schedulers are no exception! It is normally fairly difficult to predict processes and schedules to meet customer demand via the MPS because of multiple sources of process variation. Few processes are better because of the variation. To some, 3.4 allowable defects in a million opportunities might seem like a high bar, and it is. That is the logic behind the stepped performance improvement approach (Class A ERP, lean, Six Sigma). Most find that Class A is not a cakewalk and the performance criteria allow 5% noncompliance. In Six Sigma terms, that is 50,000 allowable defects in a million opportunities or approximately 3.2 sigma. Now, all of a sudden,

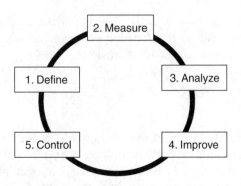

Figure 16.3. Six Sigma DMAIC Approach

a 95% accurate MPS sounds like a low objective. Class A ERP is simply a celebration point on the way to world class performance. Achieving 95% or 3.2 sigma is more difficult than you think. Process variation is as evident in the office processes as in the factory. Some of the best opportunities impacting the MPS are in areas like order management, rules of engagement, communication between the factory and the field, and strategy around new product introductions. Six Sigma has a wide berth of topics beyond the MPS itself. It really has no boundaries.

In most companies, Six Sigma is built around the DMAIC methodology (see Figure 16.3). Deming fans will see the similarities between the DMAIC process and Deming's "can-do-check-act" wheel. The approaches are based in the same learning. DMAIC would be best described as a reminder to use good process when doing continuous improvement and problem solving.

DMAIC

D: Define the Problem and the Tools to Use

One of the best tools to use in defining a problem or opportunity is called SIPOC (suppliers, inputs, process, outputs, and customers). The SIPOC is a great step to better view a process for the potential influence on improvements to it (see Figure 16.4).

It is important to know the details behind a process as the project is launched. By asking about suppliers and sources of input to the process, opportunities are opened up for solution sources that might otherwise be ignored. This is the spirit of Six Sigma — opening up all the possibilities to make sure the best solutions are found.

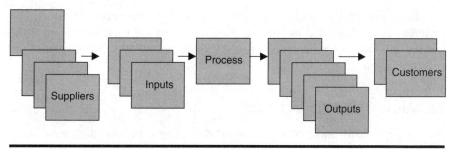

Figure 16.4. SIPOC

Outputs are just as important. When mapping the process, it is often discovered that data from the process are used by people unknown to the process owners. The reverse is also sometimes true: people supply data or outputs only to find that they are not used or needed anymore. Asking the questions posed by the SIPOC is a good exercise.

Other tools used during the Define stage include brainstorming, process mapping, and stakeholder analysis. It is easy to imagine this tool being used in a master scheduling improvement project, as master scheduling processes have many inputs (suppliers) and outputs (customers).

Stakeholder Analysis

Stakeholder analysis is another tool that can be fun for most teams. In this exercise, team members are asked to define the people who will have the most influence on a process as a proposed change is introduced. This can mean, for example, that if one member of the leadership team is very interested in this particular process and will have specific inputs and expectations, this person will be defined as a major stakeholder, even if not officially assigned to the team. Stakeholder analysis also includes defining not only level of interest but also the influence factor, both negative and positive. A client recently had to replace the vice president of operations because he had become so set in his ways that change (as determined by a continuous improvement process within the operations group) was being dragged down by his lack of engagement and enthusiasm. He was not necessarily opposing the implementation of continuous improvement, but just was not promoting it, which sent a powerful message. It is good to get this type of influence out on the table in the beginning of the project. Define who the team needs from a support standpoint and whose support the team already has. It is not always the same list of people. Ranking influence can be helpful in determining actions. Sometimes education and marketing of ideas are part of the solution.

The deliverable from this stage is an approved project scope document that includes things like what the objective is, how long the team has to finish the project, who the team members are, and even what is *not* included in or is out of scope for the project.

M: Measure and Collect Data

This stage is easily understood, but not always done with a thorough eye. There are tools that can make this stage easier and more effective. Probably one of the most widely used is process mapping. Process mapping is the documentation of a process by visibly depicting the components or actions of the process. Again, this can be very helpful in determining the opportunities and possible solutions. Several types of process mapping tools are available.

Projects in the lean space frequently use process mapping as the main tool, with design of experiment closely following. Class A tools generally are the beginning tools and often do not go beyond simple tools such as brainstorming, Pareto charts, 5-why diagramming, and fishbone analysis. In Class A, the emphasis is on establishing the cultural underpinnings of continuous improvement. Some process mapping is done during the Class A focus, but often it is limited to logical process flow type diagrams. It simply depends on the resource maturity and experience in process mapping. There is ample opportunity from the master scheduling perspective to use various process mapping tools, such as:

- **Value organization alignment mapping** — A mapping exercise to track information or decision making through an organization chart (see Figure 16.5).

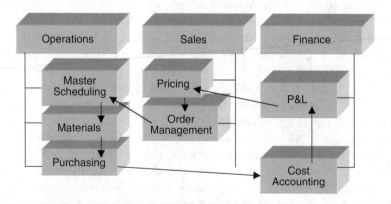

Figure 16.5. Value Organization Alignment Map

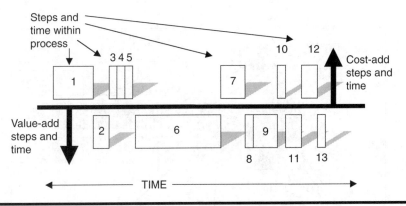

Figure 16.6. Time Value Map

- **Time value mapping** — A methodology used to map both activity and time duration of a process (see Figure 16.6). This is a unique approach in that time value becomes visible in the diagram for each step. Also, by placing value-add process elements below the timeline and cost-add activities above the line, visibility becomes a driver for change.
- **Swim lane flow charts** — Show activities separated by "lanes" of functions in the process map (see Figure 16.7). These are helpful in tracking information or material movement through different parts of the organization.
- **Physical process maps** — Process maps of building layouts showing material flow, people movement, or information flow mapped on the blueprint (see Figure 16.8).

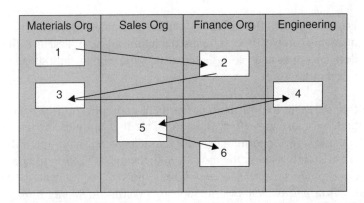

Figure 16.7. Swim Lane Flow Chart

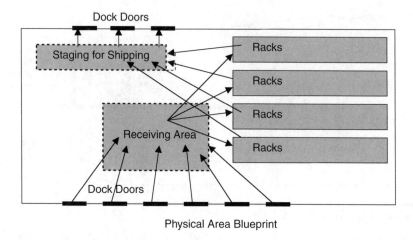

Figure 16.8. Physical Process Map

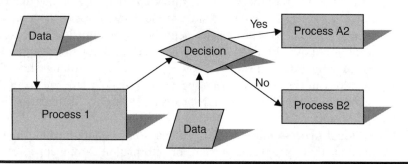

Figure 16.9. Logical Process Flow

■ **Logical process flow map** — The most common, logical process flow maps are simple process maps depicting all activities in a line, with decision points and various alternative routes shown with specific shapes (see Figure 16.9).

Another valuable tool in the measurement and collection of data stage is the fishbone or cause and effect diagram. This tool reminds the user to look in all the corners for opportunities. The reminder is normally the 6Ms: *materials* (both materials used in the process and chemicals used in the periphery), *methods* (or process approach), *machines* (tools or equipment used in the process), *measurement, manpower* (people involved), and *Mother Nature* (which can include things like barometric pressure, altitude above sea level, humidity, temperature, etc.). Obviously, any or all of these factors can contribute to process variation.

Common tools that would be used in this stage are frequency charts, Pareto charts, run charts, and metrics. The deliverables for this stage are things like a data collection plan, actual data collected, and many times a project goal validation. This stage does not necessarily end as the next stage starts. Collection of data often goes on well into and even after the project is completed.

A: Analyze the Data for Opportunities and Possible Solutions

Analysis is a key component of Six Sigma. Teams often jump to solutions too quickly. They are in the "let's try this" mode. Six Sigma requires deeper thinking and understanding in terms of causes and effects within the process in question. Some of the tools normally used in this stage are similar to the Measure stage, including fishbone analysis, process mapping, and CEDAC (cause and effect with the addition of cards). An important point in this stage is to get finance (accounting) involved in the analysis. It is especially important for the Six Sigma process to have the savings or impact verified by the financial experts in the financial aspects of the business.

The finance people should be involved in each project to a level so as to understand the potential financial impact of the project. There are different types of financial impact. The most beneficial is obviously projects that result in direct benefits that decrease existing costs and improve cash flow. There are also benefits that would be categorized as cost avoidance. These savings do not reduce budgets or improve cash flow, but can be an improvement to planned cash flow.

The Analyze stage deliverables include process maps, proposed solutions, and financial analysis reports. Financial management involvement should not be limited to the Analyze phase, but should come in any phase when there are enough data to determine a reasonable assumption of what financial benefits will come to fruition when the process is improved or the problem is solved.

I: Improve the Process

The fourth stage, Improve, is a focus on implementing the solutions determined in the Analyze stage. This does not mean that the solutions will always work. A testing process that determines the best solutions is appropriate and should be documented for future reference. The deliverables from this stage are the solutions proven and financial results verified. Celebrations are often held after the Improve stage. Measurements need to be continued well into the Improve stage to ensure that improvements have had the planned process impact. At this stage, the solutions normally have high confidence levels and are becoming the standard process, replacing prior versions.

Probably the most common tool in the Improve stage is the Gantt chart. The Gantt chart shows the actions of each expectation and the times for starting and completing various actions. In most cases, there are critical path elements that include prerequisites to some important steps. Some software is helpful in tracking these steps.

C: Control

The last step in the DMAIC process is the step to ensure sustainability. This step is normally focused on fool proofing, mistake proofing, and documentation. Documentation is especially important. Many people are proud about not wanting to take time to document and will tell others that paperwork annoys them. That is a popular view. Most people hate paperwork but have found through experience that if a process is to be enforced, the rules must be documented. It is impossible to enforce disciplines that do not exist except within people's minds. Deliverables in this space include policies, standard operating procedures, and work instructions. It is powerful to have strong documentation expectations and audits to make sure the documented process is followed, providing the documentation is in sync with the performance expectations of the business. Deliverables can also include celebrations in addition to documented savings and controlled process improvements. After implementation of the process improvements, a new sigma capability should be taken to validate that improvements were actually made.

Other Six Sigma Tools

A discussion on additional Six Sigma tools probably would start with variation-based metrics. Nothing is more important in any process improvement methodology than measurements, and within Six Sigma that is no exception. Variation-based metrics are not written as percentages. Some Six Sigma training will actually make a case against using percentages in any application. Variation-based measurements are always visibly shown with calculated limits (see Figure 16.10).

Summary

Six Sigma is a logical and rigorous methodology to help an organization reach the highest levels of process proficiency. Six Sigma is a high standard and one that can become a great motivational process. Defects are defined by the cus-

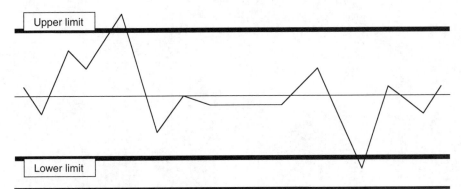

Figure 16.10. Variation-Based Metrics

tomer specification, and process improvement is determined by the Six Sigma council, normally the top management staff. It is a great tool to help with the improvement process for the master scheduling process.

This book has free materials available for download from the
Web Added Value™ Resource Center at www.jrosspub.com.

17

DISCUSSIONS WITH THE "MASTERS"

Master scheduling is not a perfect science. There are many factors that affect process. Some of these factors include the market and environment, inventory or manufacturing strategy, product complexity, the players and process owners involved, top management cooperation and ownership in the process, and of course master scheduler competence. The following interviews with some of the "masters" I have had the privilege of working with describe best practices and show how there are no perfect variation-free processes. It is hoped that something can be learned by hearing it from "the horse's mouth," as the saying goes.

You may notice that some of the practices actually deviate a little from the methodologies outlined in this book. This is the real world, which has great value in learning. The methodologies represented here are some of the best examples of good application of the ideals from this book. All are successful and profitable companies.

The three master schedulers I chose for this chapter range from relatively new in the field to experienced veterans. Their inventory strategies include (1) make-to-order forklift trucks, a major capital investment and a complex product; (2) assemble-to-order archery equipment, a consumer product with significant seasonality issues to deal with; and (3) plastic bottles, a commodity make-to-stock product with many different market issues not seen in the other products. All of these organizations are regarded as high-quality, high-performance companies in their markets.

MASTER SCHEDULER INTERVIEW #1:
BOB SHEARER, CPIM, THE RAYMOND CORPORATION

The first interview is with Bob Shearer. He has been a master scheduler at The Raymond Corporation, a high-performance, very profitable, high-end capital goods business, for over 20 years and has much to share. One of the "masters," I have learned much from Bob over several years of working with him. The Raymond Corporation (www.raymondcorp.com) is a world class manufacturer of counterbalanced narrow and very narrow aisle material handling equipment.

1. In your mind, what is the definition of the master scheduling process summed into just a few words?
Bob: "These are the definitions that come to mind quickly: supply meets demand, making black and white out of gray, reducing chaos (the disparate, often unpredictable demands of the marketplace) to arithmetic."

2. What are the most important aspects of your job as the master scheduler?
Bob: "The master scheduler's primary responsibility is to understand and reconcile supply and demand. The challenge, and what makes it interesting, is to simultaneously achieve outstanding customer service while meeting the business needs that assure profitability. It's sort of like doing a jigsaw puzzle without being able to look at the box and without knowing whether all of the pieces are there."

3. What inputs affect you the most, and what do you do to offset process variation in those inputs?
Bob: "There are two basic elements: demand, comprised of firm orders and forecasted order rates, and supply, which is the ability of manufacturing and the supply chain to meet that demand. Volatility, both forecasted and unforecasted, in demand is our major challenge. We deal with several different kinds of demand cycles — within the month, where demand peaks at the end of the period; seasonally, where demand in some months is typically higher than others; and then the overall economic cycle. This is generally 'plannable' volatility. We measure and analyze forecast accuracy and bias, do historical data extrapolations, and utilize econometric models to understand future demand as well as we can. Other volatility is harder to anticipate — buying decisions by large customers, and sometimes current events, can abruptly change your situation. The master schedule, to the extent possible, buffers the manufacturing operation from this volatility. In the face of this demand uncertainty, we keep our production schedules as stable as possible, modulating order backlog in

order to maintain level schedules. A key issue is to understand how much flexibility, in terms of your business's ability to profitably respond to unplanned demand, you actually have at any point. Continuously increasing your factory's and your suppliers' flexibility and ability to respond to unplanned demand has become a survival issue in today's competitive environment."

4. How does sales interface with the MPS process?
Bob: "The forecasting process is done within marketing at Raymond. The marketing directors develop the demand plan by product line each month and hand it off to master scheduling."

DHS comment: Marketing directors at Raymond are somewhat like product managers in other companies. They just chose a different title here. There is a sincere interest in continuous improvement in the demand plan accuracy at The Raymond Corporation.

5. How accurate is the demand plan?
Bob: "It's always wrong — but that's inevitable to some degree. The key is that it must represent the *best available* information on future demand. Marketing synthesizes distributors' forecasts, market analysis, history, and product/price/promotion info to develop its product forecasts. The toughest time to forecast accurately is when the economy is changing. Customer behavior becomes more unpredictable. Seeing the change coming, believing it, and then preparing for it are very tough calls. Our accuracy overall is in the 80 to 85% range but improving as the economic trends become a little more clear. Last year we averaged 85 to 90% forecast accuracy."

DHS comment: Raymond keeps a close watch on market demand, and the master scheduling and marketing teams work more closely than in many companies to determine the right forecast numbers.

6. Are you in the process of utilizing Six Sigma here?
Bob: "We are not now implementing Six Sigma at this time. We are, however, always looking for ways to reinforce our continuous improvement environment."

7. What is the emphasis on first-time quality or yield in this facility?
Bob: "We have a very rigorous 'first article' process with the supply chain. Suppliers are qualified and later certified by proving their trustworthiness in terms of quality and delivery. Manufacturing does first piece inspection, and additionally sample checks are done randomly throughout the process. Raymond has made its reputation on quality and innovation, so this is an area we manage closely. Quality is very rarely an issue in achieving MPS accuracy."

8. How often do the communications formally flow between master scheduling and sales? Is there a weekly or daily repeatable process review?

Bob: "We have a formal S&OP meeting once a month and continuous communications on forecast changes and special situations. In our organization, order entry reports to master scheduling, so we have a direct link to the field. This has improved the effectiveness of both the order entry and MPS functions. Over the years, order entry has reported to several different departments. It works very well here."

9. What are the system tools used by your MPS process?

Bob: "We use [a very popular ERP business system] with some bolt-on-type enhancements we designed ourselves, such as the final assembly schedule and planning bills. We assemble basically on one-piece lot orders in a final assembly schedule mode. We master schedule build slots through the use of planning bills, and as orders are received we commit the build slots and drive MRP to the customer order specifications. Most of our products are built specifically to the customer's need. That is our niche — close to a 'no two orders are alike' scenario. We also use a spreadsheet process to drive MRP requirements for some of our lower volume lines. These planning bill generators were not converted to [the very popular ERP business system] because the products are very complex with many planned variables and the spreadsheets were already in place and working. This information is directly linked into MRP."

10. What role does master scheduling play in the top management planning process here?

Bob: "If you are talking about the S&OP meeting, it's my meeting; I run it. The vice president of operations routinely attends, along with his staff, and the marketing and sales people directly responsible for the various demand forecasts. It is understood that the information has to be the best available, and process owners are expected to answer any questions about the numbers. It was started in the late 1980s when we initiated a Class A MRP II process implementation and has been evolving ever since."

DHS comment: This is a unique setting. Raymond has been at this a long time and takes the process very seriously. The company has gotten past the period of needing the president to ask tough questions. Don't be fooled into thinking top management doesn't have to be involved. Raymond's process has become part of its culture. Raymond has been doing an S&OP process well with top management support since the 1980s.

11. Is there a formal S&OP process in place? If so, what role do you play in it — preparation for it, participating in it, process ownership in it, etc.?

Bob: "I prepare the production plan, determine the major issues, distribute the information, and lead the meeting. I send out what is known as the 'stupid questions list' in advance, identifying some of the questions that will be raised at the meeting. This prework helps assure that the process owners come to the meeting prepared to address the key issues. What I call the 'stupid' questions, those that are so obvious they seem almost ridiculous to ask, are often the most difficult ones to answer. These are simple questions like 'Why are you forecasting 10 a month, when we've been selling 20 a month? What will cause demand to go down?'"

12. What roles does the master scheduler play in the management systems within your organization?
Bob: "We have daily, weekly, and monthly control processes in place. These have evolved over the past 15 years. The *daily* production meeting is held in the vice president of operations' office. It is a confirmation for the day's build and review of yesterday's performance. The meeting is efficient because people come prepared to answer any anticipated questions. Everyone works together. Shortages are not usually a big problem because schedules are based on part availability, capabilities, and resources. I attend once a week to confirm that the final assembly schedule is still valid. The *weekly* meeting is called the RAMP meeting. RAMP stands for Raymond Advanced Manufacturing Program. RAMP was our Class A MRP II certification implementation back in the late 1980s, early 1990s. This performance measurement and continuous improvement meeting has been held weekly since 1988. Process owners within the operations team report on their weekly metrics and on causes and corrective actions in cases where the result is below target. Currently, the purchasing manager leads the meeting, although several of us have had turns over the years. The vice president of operations attends this meeting. The *monthly* meeting is the S&OP meeting already discussed."

13. Do you have and attend a daily schedule review?
Bob: "As stated earlier, I attend the daily production meeting once a week. It is important to confirm the validity of the schedule based on our current status and that any issues that could affect the schedule are being addressed. This process confirms the various departments' commitment to the schedule."

   DHS comment: This once-a-week event tacked on to the daily schedule review meeting is the equivalent of the weekly Class A "clear-to-build" process.

14. Is the manufacturing process fairly consistent in demonstrated capacity?
Bob: "Yes. We have excellent, repeatable processes, a good understanding of our capacity, and we schedule accordingly."

15. Do you have a weekly clear-to-build process?

Bob: "We do not call it that, but our daily schedule review results in a clear-to-build assessment for each truck in the schedule for the next day's build. The weekly schedule review mentioned earlier confirms that the schedule is still right and supportable, from both a manufacturing and supply chain capability standpoint."

16. Do you attend a weekly performance review process?

Bob: "Yes — the RAMP meeting mentioned before. I have measurements to report, including customer delivery performance and engineering change metrics. When this meeting started several years ago, I used to facilitate it."

17. What interface do you have with the customers?

Bob: "Order entry reports to me, but I rarely communicate directly with customers. We sell through dealers, and they handle virtually all customer communication. I do get involved in unusual requests with dealers, sales, national accounts, and OEMs."

18. What rules of engagement are in place here between order management and manufacturing?

Bob: "MPS is the sole contact point for schedule commitments and questions regarding product availability. Orders are evaluated and verified before a ship commitment is made. Special orders which require extra resources are scheduled based on agreements with the groups, such as engineering, manufacturing, materials, and suppliers, that will have to do the unplanned work."

19. Is there a firm fence, fixed fence, or other lines of demarcation within the lead time?

Bob: "There is. From receipt of order to shipment, we have two lead time elements we track: (1) market required or 'target' lead time and (2) actual current lead time. The operations people here have a real sense for how important customer lead time is to our business. We balance the drive for shorter lead times against the business requirement for a stable schedule. Each product line has a target lead time and what we call a 'fill or kill' time fence. This is the signal that a slot is at the drop-dead point and we either need an order to fill the build slot or remove it from the schedule.

Our most important time fence is our rate fence, the point at which model build rates can be changed. A time fence is a warning, not a wall — it tells you to do your homework very carefully before you make a commitment inside that point. As one of our former manufacturing managers used to put it: 'If you need to shoot yourself in the foot, take the time to aim between your toes.'

Understanding and reducing this fence, by decreasing manufacturing flow times and decreasing purchased part lead times, is the key to improving your flexibility in responding to volatility in customer demand."

20. Is top management aware of these rules? Are they enforced? By whom?
Bob: "Yes. Our management has a good sense of these rules and what they mean. The rules are enforced by MPS. There are really no issues here with what you call 'rules of engagement.'"

21. What is your role in new product introduction?
Bob: "Change management also reports to me. This group is responsible for the coordination of engineering changes for existing and new products. New product introduction is done through a cross-functional team led by a program manager. Master scheduling is involved in each introduction — we are responsible for the scheduling of the pilot build and for the phase-in/phase-out plans at the model and part levels. We are doing an increasing number of new product introductions and are finally getting pretty good at it."

22. Does your new product introduction process have measured deliverables and gates regularly reviewed?
Bob: "We use the standard six-phase product development process. Management formally reviews the deliverables from each involved group at each phase."

23. Which inventory strategies do you use here?
Bob: "We use them all. Most of our products are built in the ATO [assemble-to-order] mode, but some are MTO [make to order] or ETO [engineer to order]. The least used inventory strategy in our world is the MTS [make to stock]. We sell what the customer really needs based on a detailed analysis of the application, not what we happen to have on hand. As a result, there are only a few configurations of a few models that are standard enough to stock. In these more popular configurations, dealers sometimes stock units also, giving us a buffer between customer demand and the factory."

24. How many product families are there?
Bob: "We build 10 product families and 25 individual models at this location."

25. How many planners are there in the material department supporting this operation?
Bob: "We have four planners supporting production, four supporting the supply chain, four supporting ECNs and new products, and three supporting the master scheduling operations."

26. Do you use planners/buyers or are the processes separated?
Bob: "We use buyer/planners."

27. How many items are typically planned by each planner?
Bob: "Each planner handles about 4,400 total part numbers, but not all are active at the same time."

28. Are there any items on autopilot, where no one looks at the orders prior to releasing them to the supply chain?
Bob: "Almost all manufactured parts are on autopilot. They are being driven by a firm customer order by the time they are released to the shop. Welding and some of the cells in fabrication work from a pull system from the MPS. On some of our higher volume products, we regularly take product from raw plate steel to painted lift truck in six workdays. Purchased parts are not on autopilot except for commodities such as hardware and decals. Those are controlled by a vendor-managed inventory process. Many of the critical purchased items such as steel uprights and transmissions are scheduled for daily just-in-time delivery and sequenced to match the final assembly schedule."

29. Do you use ATP [available to promise]?
Bob: "We use ATP at the final assembly schedule build slots level. The availability of open build slots dictates the ship date of an order. Because we level load the schedule as much as possible, changes in demand can cause lead time for a specific product to fluctuate."

DHS comment: Raymond has a good ATP process. I subscribe to the philosophy that there is no such thing as "standard lead time." Lead time can fluctuate as spikes in demand happen. As an old boss of mine used to say, customers are usually ill-behaved.

30. Does order management execute the ATP or does the master scheduling department?
Bob: "Master scheduling assigns the final assembly scheduled slot and ship date."

31. How many items per person do you master schedule in your organization?
Bob: "From about 300 to several thousand. The key factor in our process is not the number of part numbers, but the volume of sales orders to be scheduled."

32. What does your organization look like?
Bob: "I report to the materials manager, who reports to the vice president of operations. Three functional areas report to me; MPS, order entry, and engineering change implementation."

33. Is the demand plan published in the same product families that you break schedules into?

Bob: "If you mean does the whole business plan around the same product families, the answer is yes. The S&OP process including the demand plans, production plans and the MPS are all categorized in the same product line designations."

34. Is the demand plan measured, and are people held accountable for improvements or root cause?

Bob: "The forecast is measured, and the marketing product directors as well as the vice president of marketing are all very determined to improve the performance. There is, however, no formal process of reporting root cause of these sources of variation when the forecast is incorrect."

35. What are the performance measurements used here that support the processes in master scheduling?

Bob: "All the standard Class A ERP–type metrics are in place. At our weekly performance review meeting we measure about 60 aspects of the business, such as delivery performance, weekly build rate attainment, daily schedule adherence, on-time work orders, percentage of pickable work orders, supplier delivery, supplier quality, engineering change turnaround time, inventory accuracy, open work order status, data accuracy, and production machine uptime. Most of our measures are consistently at the Class A level."

36. Who are the process owners for those measurements?

Bob: "The managers of the areas are normally the process owners — they're the ones responsible to make their process work effectively. For example, as manager and process owner of MPS, I make the policy decisions, lead the negotiations in major scheduling situations, make sure that the rules are followed, and continuously look for process holes and improvements. Process ownership is taken very seriously here."

37. Is data accuracy disciplined here adequately?

Bob: "We have spent a lot of time and effort on data accuracy, but it's never really conquered. It has not been really out of control for years, but could always be better, to be honest. Our work-in-process accuracy is the biggest challenge. Our stockroom balance accuracy is normally in the mid to high 90s calculated by location. Work-in-process location accuracy runs slightly lower than that, and we continue to focus on root causes. Inventory accuracy has not seriously affected our MPS accuracy for many years."

38. What is the inventory accuracy in the warehouse here?
Bob: "High 90s."

39. What is the customer delivery performance from this facility? How do you measure it?
Bob: "It has been between 97 and 100 for many years. It is 100% most of the time. That is mainly because of the consistent manufacturing process and the disciplines around scheduling to the capability. Delivery is measured to original customer promise."

40. What are the product lead times from receipt of order to shipment of product on the majority of products from this facility?
Bob: "Our customer lead times vary from product to product but can be as short as 1 week and as long as 20 weeks. Most are in the 9- to 10-week range for ATO products."

    DHS comment: Unlike some markets, the custom material handling business is a little more forgiving on lead time. While 20 weeks may seem like an eternity to some readers, it is market driven and is under constant scrutiny at Raymond for shortening. It has been improved significantly on most high-volume lines.

41. Does engineering play a role in the master scheduling process?
Bob: "All orders are run through an order configurator. If they get flagged as special for any reason, engineering is required to review the order and provide the required custom design. This design time is factored into the scheduling of the order."

42. Approximately how many engineering changes a week do you see?
Bob: "About 60."

43. What would you say to a new person entering the field of master scheduling as advice? Any words of wisdom?
Bob: "To be an effective master scheduler, someone should learn as much about their customers, suppliers, manufacturing processes, and cost accounting as they can, in addition to the mechanics and theory of the master scheduling process itself. I worked in manufacturing, human resources, and later in marketing before I came to master scheduling. I think a varied background is a major advantage. It also helps to have a thick skin, the ability to attack and solve problems logically, and the ability to identify creative alternatives and to say no when necessary. Raymond takes master scheduling very seriously and pro-

vides the authority and control to do it right. I have had many different jobs, and this one is the most fun."

DHS *comment:* Bob has been instrumental in developing the robust master scheduling process at this facility. It is one of the reasons why Raymond is one of the best-managed manufacturing processes I have seen. Hats off to Bob and his team!

MASTER SCHEDULER INTERVIEW #2:
JEREMY SAUER, CPIM, MATHEWS, INC.

Jeremy Sauer has fewer years of experience than the other two masters chosen for this chapter's interviews. He has nonetheless built a solid MPS process and one with special focus on manpower capabilities and accuracy in scheduling. It is an impressive show of discipline and linkage between the sales team and the plant. Mathews, Inc. (www.mathewsinc.com) changes inventory strategies at varying times depending on where it is within the demand season. This organization also demonstrates good creativity with planning bills. As would be expected with that kind of success, top management is very supportive of the measures and management systems that drive improvements in this business. Mathews, Inc. has world-class products and is working to maintain world-class processes. It is generally thought of as the leading manufacturer of high-performance archery equipment. Jeremy Sauer and the others in the materials group are dedicated to keeping their process in the lead. He recently organized the MPS process to allow Class A ERP certification. The Mathews logo phrase — "catch us if you can" — seems very appropriate for this team.

1. In your mind, what is the definition of the master scheduling process summed into just a few words?
Jeremy: "Master scheduling process requires using industry knowledge, analytical and technological capabilities, and cumulative experiences to schedule a manufacturing strategy that repeatedly meets short- and long-term demand in a proficient system."

2. What are the most important aspects of your job as the master scheduler?
Jeremy: "The most important aspect is to build a reputation for Mathews, Inc. through schedule adherence and good promising that is consistent and exceeds customer delivery expectations. As the master scheduler, I can help tremendously by planning material and people cost effectively to return the greatest value for the customers and Mathews, Inc."

3. What inputs affect you the most, and what do you do to offset process variation in those inputs?
Jeremy: "The most important input is the demand plan. The demand plan outlines units of sales by product family. This is critical when scheduling using planning bills of material. There is some variation introduced into the process from the demand plan. While our sales people do a great job, it is difficult to always have it close.

"I offset forecast process variation by understanding the variation itself. If it is variation in sales at the product component level, then I use a planning bill of material to create small amounts of safety stock at the item level. I audit the planning bill of material component ratio percentages monthly to reduce the risk of misalignment with reality. This helps maintain the desired safety stock level.

"To cover variation in units of sales at the family level, I can use safety stock at the planning bill of material level to buffer demand variances. Our sales team calls this variation in sales a 'stretch plan' or an optimistic view of sales for the period.

"In an MTS approach, I plan for the end the month to equal the inventory plan. A small variation within the family has very little impact, because a planning bill of material audit and MRP will correct it over time.

"Offsetting variation created by manufacturing process capabilities requires eliminating the problem through TQM techniques. In some cases, we use yield percentages to protect inventory levels."

4. How does sales interface with the MPS process?
Jeremy: "The sales team identifies root cause of trends in consumer ordering. The team also plans for future promotions and features. They communicate with us frequently. The sales team is very interested in keeping the MPS accurate, allowing for the best customer service."

5. How accurate is the demand plan?
Jeremy: "The demand plan on average is about 95% accurate at the aggregate family level. The family's unit level is approximately 85% accurate. The sales team is constantly looking at this for improvement."

6. Are you in the process of utilizing Six Sigma here?
Jeremy: "We currently do not use Six Sigma techniques. Quality is a strength of Mathews, Inc., however. We know the importance of quality control, and we are documenting and measuring quality opportunities such as mean time between failures. Six Sigma concepts would probably be valuable for Mathews."

7. What is the emphasis on first-time quality or yield in this facility?

Jeremy: "We are learning to be more proactive in the areas of yield. We do have a material review board committee that measures the impact of quality. We will be adding more quality measures in the upcoming year. We are moving into an emphasis on lean and waste elimination. Yield or first-time quality issues are an opportunity to eliminate more waste."

8. How often do the communications formally flow between master scheduling and sales? Is there a weekly or daily repeatable process review?

Jeremy: "Master scheduling sends out a daily report with stock for the limited MTS we do and ATP information for MTO and ATO products to the sales and customer service teams. We meet weekly to discuss sales variations, stock issues, and anticipated forecasts with the sales team."

9. What are the system tools used by your MPS process?

Jeremy: "Made2Manage software for the MPS Calendar, Microsoft Query, planning bill of material development in M2M, and Microsoft Excel."

10. What role does master scheduling play in the top management planning process here?

Jeremy: "The master scheduler converts the demand plan into a 12-month rolling operations plan. A capacity requirements plan is then developed. The capacity requirements plan tells us the staffing levels each month and what impact the demand plan will have on human resource management: need for temporary services, permanent full time, permanent part time, seasonal employment, etc. The capacity requirements plan also tells us machinery requirements in each department in the short- and long-term planning horizon."

DHS comment: Mathews, Inc. has an excellent process to schedule resource requirements. The company makes decisions to move people based on its MPS week to week.

11. Is there a formal S&OP process in place? If so, what role do you play in it — preparation for it, participating in it, process ownership in it, etc.?

Jeremy: "We call it SIOP [sales, inventory, and operations planning] rather than S&OP. One of our top managers used to work at AlliedSignal, where they called it SIOP. Taking the 12-month rolling demand plan, I develop a 12-month operations plan. As the process owner for the operations plan, I develop it using a manufacturing strategy [also known as inventory strategy — MTS, ATO, etc.] that best uses our resources to deliver product on time. I am responsible for the monthly measurement and developing an MPS with weekly clear-to-build numbers to meet the monthly operations plan. The director of operations is the owner of the SIOP and facilitates the meeting. It is held predictably every month and is an important part of the management system here."

12. What roles does the master scheduler play in the management systems within your organization?
Jeremy: "I determine the manufacturing strategy [also known as inventory strategy] and desired inventory levels. We are a fairly vertically integrated company owning almost all processes in-house. This helps our quality control and flexibility but creates extra scheduling alignment requirements for scheduling. Master scheduling needs to utilize and direct a cross-trained workforce effectively."

13. Do you have and attend a daily schedule review?
Jeremy: "We eliminated the daily review and instead just have a weekly schedule review. We found that a daily review is too redundant because our scheduling process has become very accurate using ATP. The weekly review outlines department production schedules for the next five days. All component details are executed through job order creation and dispatch reports. This is also referred to as the clear-to-build process."

14. Is the manufacturing process fairly consistent in demonstrated capacity?
Jeremy: "Yes. Our demonstrated capacity is very consistent, although we have some seasonal demand cycles that affect this predictability. When we have a process change, we reevaluate our production rates. At certain times of the year, we change from an MTS to single piece flow (MTO) strategy. This can affect rate, and demonstrated capacity is adjusted to the new rate capability. The trade-off for less efficiency is greater inventory turnover with timely supply depending on the season."

15. Do you have a weekly clear-to-build process?
Jeremy: "Yes. The schedule is outlined each week. Each manager knows the production expectations prior to the week starting. The results are measured and reviewed weekly."

16. Do you attend a weekly performance review process?
Jeremy: "Yes and no. We always know our performance to the MPS weekly, but if it is at or above plan, we don't usually discuss it at the weekly review. Not to take any emphasis off the review meeting importance, however. As a

colleague of mine once said, 'we run our business by the numbers.' We understand our success when we are meeting plans. We use the data as a gauge and to drive actions. When there is process variation, we do discuss it at the weekly review. We use the time otherwise to improve process, such as directing small-group improvement activities problem-solving effort."

17. What interface do you have with the customers?

Jeremy: "I interact with customers indirectly through the weekly sales meeting. When auditing the planning bill of material, I can see the variation in customers' choices. The sales team and I try to determine if this variation is special case or normal random process noise."

18. What rules of engagement are in place here between order management and manufacturing?

Jeremy: "Our goal is to work to daily schedules. This gives us a clear objective to accomplish daily. Our scheduled buckets for manufacturing include lot sizes with operations that can be accomplished within one day. Our purchasing policy has goals to maintain either a fixed quantity or days supply on-hand inventory that is compatible with supplier lead times."

19. Is there a firm fence, fixed fence, or other lines of demarcation within the lead time?

Jeremy: "Each product family has a different planning horizon. The MPS planning time fence is time phased by each planning bill of material to cover the demand for the longest lead time component. The MPS is loaded in weekly buckets starting with the beginning of every work week. We use a backwards scheduling technique from the date of customer orders. Our ATO strategy allows us to ship to customers within three days. Customer requests that are less than three days are often accommodated, but must be approved by the production planner to make sure it is a promise we can keep."

20. Is top management aware of these rules? Are they enforced? By whom?

Jeremy: "Top management is aware and entrusts us to set these parameters and implement them across the supply chain. The rules get reviewed through a performance metric called 'Materials Plan.' This metric compares the job order calculated start date and the job order release date. It also measures if purchase order receipt dates are prior to the job order needed date and if job completion is on time to the job order due date. If performance is not tracking to goals, root cause is investigated."

21. What is your role in new product introduction?
Jeremy: "I include new product in the operations plan. The buyer begins ordering approved raw material before the final design. Raw material is our longest lead time component. Once new product designs are finalized, I create standard routings to match new processes when applicable."

22. Does your new product introduction process have measured deliverables and gates regularly reviewed?
Jeremy: "Yes and no. New designs and technology are a Mathews competitive advantage. We are flexible with engineering change orders and development dates. However, we do set gates for bill of material implementation and production start-up. We also measure delivery of new product to the customer after designs are finalized. We schedule and build to deliver a standard preview bow for every new product to all of our retailers within the same week of introduction."

23. Which inventory strategies do you use here?
Jeremy: "MTS, ATO, MTO, and ETO. We also occasionally use ETO to fit special dimensions, such as oversized bows made for large athletes. We are a seasonal business, so our manufacturing [inventory] strategies change throughout the year."

24. How many product families are there?
Jeremy: "We currently have six product families."

25. How many planners are there in the material department supporting this operation?
Jeremy: "We have two planners, one responsible for manufactured items and one responsible for purchased items."

26. Do you use planners/buyers or are the processes separated?
Jeremy: "Yes, we have a planner/buyer. Our purchasing planner answers signals from MRP and releases purchase orders."

27. How many items are typically planned by each planner?
Jeremy: "Each planner typically plans 600 parts."

28. Are there any items on autopilot, where no one looks at the orders prior to releasing them to the supply chain?
Jeremy: "Some hardware items are on autopilot, but most inventory items are reviewed by planners prior to release."

29. Do you use ATP?

Jeremy: "We absolutely use ATP; it keeps us accountable to our customer. We have a rule-based system that uses different ATP measurements during periods of backlog and throughout our business cycle of one year. The first quarter is often our new product introduction period. ATP is stated in the week of a month. In the second quarter, we specify an explicit date. The third quarter is heavy for us and has a standard lead time of three days for shipment from order placement. If requested, we can ship MTS items within one day. The fourth quarter has a three-day standard shipment from order placement. We do much less MTS in the fourth quarter as the season is winding down and we are preparing for new product introduction stock shifts."

30. Does order management execute the ATP or does the master scheduling department?

Jeremy: "The MPS department sets the ATP. We utilize the order management software to maintain due date priority, but the planner must determine if we have the capabilities to meet the demand (capacity/material). The planner manually converts MTO requirements into job orders."

31. How many items per person do you master schedule in your organization?

Jeremy: "We have only one master schedule. I master schedule 70 different planning bills of material and 40 individual configurations. Together this represents approximately 1,200 items."

32. What does your organization look like?

Jeremy: "We have a unique flat organizational structure. Everybody reports to our president in an open-door atmosphere. We have a matrix reporting structure for process functions. For example, I also report indirectly to the materials manager for process decisions."

33. Is the demand plan published in the same product families that you break schedules into?

Jeremy: "Yes. This makes both the demand plan and MPS metrics significant when comparing month-end performance."

34. Is the demand plan measured, and are people held accountable for improvements or root cause?

Jeremy: "Yes. We measure the demand plan monthly and sometimes, depending on the season, bimonthly. Adjustments are made, and root cause is determined."

35. What are the performance measurements used here that support the processes in master scheduling?
Jeremy: "We measure the execution of all weekly departmental production goals from the on-time completion of all job orders throughout the process."

36. Who are the process owners for those measurements?
Jeremy: "Department managers and ultimately each associate at Mathews, Inc. is part of the measurements."

37. Is data accuracy disciplined here adequately?
Jeremy: "This is always a continuing process improvement when new items are introduced. We measure our accuracy level and perform root cause analysis. We are developing a cycle count procedure in work in process that necessitates all associates' participation. Fail-safe tools are continually being implemented to alleviate inaccuracies."

38. What is the inventory accuracy in the warehouse here?
Jeremy: "In controlled locations where picking takes place, inventory accuracy is 95 to 96% accurate in terms of pieces per location. We are continuing to work on our point-of-use accuracy and have recently implemented procedures to improve our accuracy on the shop floor. This requires timely transactions when items are moved from one area to another."

39. What is the customer delivery performance from this facility? How do you measure it?
Jeremy: "We measure using the sales order due date versus date shipped. We are almost always over 95%. We manage a backlog during new product introduction, but ATP is understood by the customer."

40. What are the product lead times from receipt of order to shipment of product on the majority of products from this facility?
Jeremy: "Three days."

41. Does engineering play a role in the master scheduling process?
Jeremy: "Engineering always plays a role in master scheduling. New planning bill of material and inventory strategies are created during new product introduction. The sales team also suggests what models should be phased out and what common components to use in new models. My job is to lead the strategy to reduce obsolete inventory and deliver new product to the market quickly."

42. Approximately how many engineering changes a week do you see?
Jeremy: "We see very few ECNs a year. Most of these are associated with new products. Once the product is in production, it has a mature design. This is due to the testing done prior to launch."

43. What would you say to a new person entering the field of master scheduling as advice? Any words of wisdom?
Jeremy: "Continuing education contributes to the success of a master scheduler. Pursuing APICS certification in CPIM/CIRM has been my most rewarding experience of learning in this field. You must have a theoretical foundation to see the impact of your decisions. This field also requires advanced analytical skills. Skills can be developed through experience, but you must have the theoretical knowledge to ask the right questions.

"Master scheduling is a very rewarding profession if you like setting goals and accomplishing goals with the combined strengths of others. It is a profession that requires self-teaching and also learning from the experience of others. The best learning activities can take place while benchmarking outside your own plant. Another good source of learning can be found at the on-line bookstore www.apics.org."

DHS comment: APICS recently limited its bookstore offerings to books referenced by its certification process documentation. Amazon.com is also a good source for materials associated with master scheduling.

MASTER SCHEDULER INTERVIEW #3: ROBERT TURCEA, GRAFCO PET PACKAGING COMPANY, INC.

Grafco PET Packaging (www.grafcopet.com) is a blow-molding company that serves several food and pharmaceutical markets with PET packaging solutions. It has plants in several states and continues to grow. The interesting thing about this master scheduling environment is that all products and lines are master scheduled from the headquarters in Baltimore, where Robert Turcea resides. There are no master schedulers in the remote plants. The rules of engagement are well understood, and the sales team is very active in the top management planning process. Robert Turcea has been in this position for a few years. He has built a reputation internally for discipline and keeps up with this important job very nicely.

1. In your mind, what is the definition of the master scheduling process summed into just a few words?
Robert: "The coordination of demand, forecast, and production to meet customer requirements and minimize inventory holdings."

2. What are the most important aspects of your job as the master scheduler?
Robert: "Communication with the plants and customer service, as well as constant attention to trends and forecast changes from the demand team. I am on the phone with the plant contacts all day long or talking with the customer service people. Communication is extremely important for our successful coordination of plant activity for customer need."

3. What inputs affect you the most, and what do you do to offset process variation in those inputs?
Robert: "Significant unplanned demand spikes without sufficient lead time affect scheduling accuracy the most. I analyze safety stock levels as frequently as possible and work with demand planning to use history as a tool to smooth the loads to operations. We have to serve customer needs, but we also have to keep costs in line."

4. How does sales interface with the MPS process?
Robert: "The demand manager, who reports to the vice president of sales and marketing, sits 10 feet from me in the same walled office. We talk constantly. The demand manager understands master scheduling and the issues surrounding it because he used to do it. The demand manager uses Demand Solutions software, and through the monthly demand cycle of inputs and analysis we follow procedures that allow the handling of schedule changes and specific customer requirements. We work together pretty well, which makes the master scheduling job much more successful."

5. How accurate is the demand plan?
Robert: "Currently the demand plan runs about 75% accuracy. We measure by product family and tool. We have a pretty tough metric, but top management is very supportive. The accuracy is improving, and the vice president of sales and marketing is very involved in the causes for variation."

6. Are you in the process of utilizing Six Sigma here?
Robert: "Not yet, but it is definitely on the horizon. We are in the middle of a continuous improvement process [CIP] implementation which has both a lean

and Six Sigma flavor. We have a process where project leaders can be certified as CIP green belts and CIP black belts, but it is more of a Six Sigma 'light' at this point."

7. What is the emphasis on first-time quality or yield in this facility?
Robert: "There is a strong emphasis on first-bottle quality on each setup involving the quality manager, production manager, and maintenance manager, and no product is scanned into inventory until all three have been approved. We are also in the process of making our cost-of-quality metric more encompassing to capture all costs and drive improvement. Quality is not normally a problem for the master scheduling process at Grafco."

8. How often do the communications formally flow between master scheduling and sales? Is there a weekly or daily repeatable process review?
Robert: "The communications flow is constant on specific orders of item numbers. For the product families, there is a more formal monthly S&OP process which is closely tracked. There is a monthly demand review process, but because the demand manager and I share an office and sit next to each other, we communicate continuously. It is a very good arrangement."

9. What are the system tools used by your MPS process?
Robert: "We implemented both Demand Solutions software and CMS ERP software a few years ago when we first started our Class A implementation. Demand Solutions is our demand manager tool, and master scheduling is within the CMS ERP business system software package. Both work well."

10. What role does master scheduling play in the top management planning process here?
Robert: "Master scheduling's role is mostly to provide analysis of capacity issues and tooling constraints on which top management bases some decisions about new customer demand focus and product family capacity commitments to the marketplace."

11. Is there a formal S&OP process in place? If so, what role do you play in it — preparation for it, participating in it, process ownership in it, etc.?
Robert: "Yes. I set the production requirements based on the demand plan for the rolling 12 months and attend the production phase of the S&OP meeting cycle. I also track progress to the plan weekly and measure the plant run hours daily. I also review all major changes to machine capability and tool capacity. This includes changeover schedules and issues with tooling, etc."

12. What roles does the master scheduler play in the management systems within your organization?
Robert: "My job is to keep schedules realistic and doable. The plants actually have process ownership for the schedule adherence, but if I schedule unrealistic or unlikely plans, I hear about it. It really isn't an issue. I do not attend a weekly performance-review-type management system event. I primarily affect the management of the organization from a coordination standpoint between the demand process, the customer service team, and the manufacturing facilities."

13. Do you have and attend a daily schedule review?
Robert: "No. The formal schedule review at the corporate level is a weekly meeting. The daily attention comes from my individual review of the production progress in each plant and a schedule review at each plant attended by the plant managers and supervisors. I do not attend these, as there are several happening at each plant simultaneously each morning."

14. Is the manufacturing process fairly consistent in demonstrated capacity?
Robert: "Within established or mature product lines, demonstrated capacity is very consistent; however we are in a strong growth period involving many new products. There is much less consistency with new tools, products, and machines. We have to be much more cautious with these."

15. Do you have a weekly clear-to-build process?
Robert: "Yes. We have a formal schedule review process each week. Plants are given the opportunity to accept or give feedback on the schedule for next week. Most of the time, there are few changes because I am so close to the capacity issues on a day-to-day basis anyway. This schedule review is just a formal process to make sure the plants have a chance to affect the schedule. They own it from then on."

16. Do you attend a weekly performance review process?
Robert: "No. However I do review and update metrics which are used in the plant meetings as well as the corporate meeting. The weekly operations review meeting is a conference call review with the plants that is held each week. I do not attend this; the director of materials [to whom Robert reports] does however."

17. What interface do you have with the customers?
Robert: "I have almost no direct contact with external customers. My interface is with the customer service team and occasionally the outside sales team."

18. What rules of engagement are in place here between order management and manufacturing?
Robert: "We have procedures about minimum runs and lead times for MTO products, as well as a procedure in development to address schedule changes on a case-by-case basis. This is an important part of our master scheduling success here. Management respects the rules, and the demand-side people have the responsibility to maintain the rules if they are not working."

19. Is there a firm fence, fixed fence, or other lines of demarcation within the lead time?
Robert: "We have traditionally had a 10-day time fence on production schedule changes. However, we are in an evolutionary process that is concentrating more on the feasibility and impact of schedule changes as the determining factor as opposed to strict calendar day rules. We want to have more flexibility, and the plants are concentrating on setup reduction to increase that factor."

DHS comment: Remember that this is a mostly MTS environment. The firm fence is on high-demand spikes where buffers were not adequate and short lead time MTO requirements, which represent a small part of the business.

20. Is top management aware of these rules? Are they enforced? By whom?
Robert: "Top management is very aware of these rules. Admittedly, there are more exceptions than our original intent would support, but when tested, management supports the right decisions. We are working on improving consistent application of the rules to the extent necessary to maintain minimum plant disruptions. Sometimes customers just don't do what we planned! Forecasting is not an easy thing to get right all the time."

21. What is your role in new product introduction?
Robert: "I schedule sampling and coordinate first production runs along with the customer service team and engineering. I also analyze tool and machine capacity for the quoting process when requested. I am quite involved and have a significant role in determining the schedules."

22. Does your new product introduction process have measured deliverables and gates regularly reviewed?
Robert: "Yes. The projects are managed by a project manager who is usually an engineer. Reviews are held for each 'gate.'"

23. Which inventory strategies do you use here?
Robert: "We use both MTS strategy and MTO strategy. The great majority of our business is MTS. Many times, our customers give us only a few hours notice to ship product."

24. How many product families are there?
Robert: "We have 130 product families."

DHS comment: Grafco has product families built around tools, which are where its constraint issues reside. One tool, with modifications, can be used for more than one product. At another level, Grafco looks at machine groupings, which are much less than the 130 "tool" families.

25. How many planners are there in the material department supporting this operation?
Robert: "There are two planners at the corporate level where I reside. In addition to my master scheduling duties, I am responsible for planning in some of the plants. The other planner works with the balance, as well as many other corporate buying duties."

26. Do you use planners/buyers or are the processes separated?
Robert: "Because of our very shallow bills of material and remote plant locations all over the United States, material purchasing to support production, such as maintenance materials, machine wearable components, or MRO items, happens at the plant level, with the exception of first runs on new tools or corrugated materials."

27. How many items are typically planned by each planner?
Robert: "On average, planners have about 250 items, including both finished goods and other items at the corporate level. It is much less at the plant level. Planners at the plant level also are involved in warehouse data management and shipping normally."

28. Are there any items on autopilot, where no one looks at the orders prior to releasing them to the supply chain?
Robert: "No, not really. We have complications caused by lots of new products and tool-sharing issues between facilities."

29. Do you use ATP?
Robert: "We do not fully utilize the ATP capabilities of our system at this point. However, there is currently a strong push to use this hand in hand with finite scheduling modules in CMS and order entry guidelines from the customer service department. When we get this fully implemented, it will be a great step forward in our focus on continuous improvement."

30. Does order management execute the ATP or does the master scheduling department?

Robert: "We promise out of stock but do not use the ATP feature as designed in our software. Customer service manages this process."

31. How many items per person do you master schedule in your organization?

Robert: "Around 100 items."

32. What does your organization look like?

Robert: "We have a very thin, flat organization both at corporate and at the plant level, with a great deal of multitasking for each position. I report to the director of materials, as do the other materials people."

33. Is the demand plan published in the same product families that you break schedules into?

Robert: "Yes. It must be to be meaningful to the supply side of the business."

34. Is the demand plan measured, and are people held accountable for improvements or root cause?

Robert: "Yes. Forecast accuracy is measured and people held accountable. Our CEO takes a special interest in this accuracy and understanding root cause of surprises."

35. What are the performance measurements used here that support the processes in master scheduling?

Robert: "There are various measurements used, including but not limited to forecast accuracy, production plan accuracy, inventory goal analysis, tool change hour accuracy, and delivery performance."

36. Who are the process owners for those measurements?

Robert: "Forecast and demand plan accuracy are owned by the demand manager; production plan accuracy and inventory accuracy are held by the plant managers and me jointly. Delivery performance is owned by customer service representatives, but each plant manager and I are both closely involved, obviously."

37. Is data accuracy disciplined here adequately?

Robert: "Yes. We do a good job of verifying data before it is used to analyze, as well as questioning data that seem suspect. Generally we have accurate data to work with."

38. What is the inventory accuracy in the warehouse here?
Robert: "Inventory accuracy is typically above 95% measured as percent of perfect location balances."

39. What is the customer delivery performance from this facility? How do you measure it?
Robert: "Typically, the delivery performance is >98%. The measurement is based on the number of order line items shipped complete by the promise date."

40. What are the product lead times from receipt of order to shipment of product on the majority of products from this facility?
Robert: "For items that are MTO, the lead time is typically three to four weeks primarily due to shared tooling and predetermined campaigns. It can be much shorter. As is usually the case with multiple product lines with shared tools, there are many, many exceptions to this rule."

41. Does engineering play a role in the master scheduling process?
Robert: "Yes, from the perspective of new product introductions, major machine/tool repairs, new machine introductions, and product samplings."

42. Approximately how many engineering changes a week do you see?
Robert: "On average, there are four engineering changes per week, mostly related to carton changes or cycle times."

43. What would you say to a new person entering the field of master scheduling as advice? Any words of wisdom?
Robert: "I would recommend that a person entering this field get a strong background in actual production floor management and customer service first and also certify in APICS training. The next step would be at least some basic exposure to the machinery in their particular product line and then training in demand management and software applications. They should be prepared because they will be the hub of the supply chain wheel. Experience keeps this hub from 'squeaking.'"

SUMMARY

Although these three master schedulers work in different environments, you can see similarities in their practices. All have very disciplined cultures where the MPS is the hard-and-fast rule. All three manage change to a minimum requirement and yet have very happy customers. Some of their real-world practices deviate from the descriptions in this book yet are sanctioned by the author. Simply stated, this means that there are exceptions to every rule. The requirement is that a handshake be in place when these practices are agreed upon. The bottom line is good customer service with minimum cost. That is what we all seek.

Master scheduling is the heartbeat of every manufacturing process. It is where some of the most important decisions are made concerning the service levels and costs of manufacturing. People who choose this line of work and excel at it are special people, indeed! It is the goal of this book to cover the aspects and idiosyncrasies in enough detail that you can improve the MPS process at your company and achieve similar high service levels with low costs. Godspeed!

Web
Added
Value™

This book has free materials available for download from the
Web Added Value™ Resource Center at www.jrosspub.com.

GLOSSARY OF TERMS

ABC classification — Stratification strategy for dividing inventory into categories of importance. It is used in planning and other materials function decisions.

APICS (American Production and Inventory Control Society) — Professional society for materials and operations professionals worldwide.

Assemble to order (ATO) — Inventory strategy where inventory is kept one level down from the final customer ready and assembled as customer order is received.

Available to promise (ATP) — Calculation in which the dates are determined for available planned inventory in the master production schedule.

Backflush — Transaction automatically triggered by another transaction. Usually found in inventory balance transactions. Also sometimes called auto-deduct.

Balance by location — An inventory balance for only one location of a material or item.

Barometric measurements — Measures that apply to all manufacturing businesses. These measurements are normally ongoing for monitoring purposes. *See* Diagnostic measurements.

Bill of material (BOM) — Documented description of components and ingredients, including quantity per and item/material numbers.

Bill of resource — Record within an enterprise resource planning business system that includes both routing and bill of material information.

Business imperatives — Top priority goals or objectives that are to be accomplished in the next 12 months. Normally a short list of the most important objectives to support the strategic plan.

Business plan — Top management plan with strategic objectives, financial objectives, and business imperatives.

Business system — A computer system used to communicate linkages between processes such as scheduling and execution of schedules, inventory consumption and balance records, forecasting and planning inventory, financials and operations, etc. Usually referred to as an enterprise resource planning system.

Class A ERP — A specific high level of performance using enterprise resource planning methodology and defined by measurements and certification criteria. Organizations (such as Buker, Inc.) certify this performance level by auditing performance.

Class A ERP certification — Meeting criteria as defined by the Class A ERP requirements for process design, management systems, and results.

Clear-to-build — Meeting held at the end of the week to achieve accountability for the upcoming week's production. Normally led by the master scheduler.

Cost standards — Cost records per stockkeeping unit stored in the item master of the enterprise resource planning business system. Inventory is normally valued using these standards.

Cycle count — Scheduling and periodically auditing inventory balances for accuracy.

Daily schedule alignment — Process of realigning the master production schedule each day for any process variation experienced.

Daily schedule review — *See* Daily walk-around.

Daily walk-around — Schedule and quality review done each morning by management in a manufacturing facility.

Data fields — Areas in a database where specific information is to be stored. These fields can be user defined or system defined.

Data mining — Investigation using data. Massaging databases to help with problem solving.

Demand manager — Person responsible for developing the demand plan process with the demand side of the organization and feeding it to operations.

Demand plan — Forecast of customer demand usually produced by the sales and/or marketing department for use by operations. Inputs include business plans, marketing plans, sales plans, and history.

Demand review — Meeting between the master production scheduling people and the demand people for the purpose of reviewing demand accuracy.

Demonstrated capacity — Capacity as demonstrated in the last few process runs. Not to be confused with theoretical capacity.

Dependent demand — Demand driven from another parent requirement in the system.

Diagnostic measurements — Measures that are specific to a problem-solving endeavor. Used for fact finding. These are sometimes monitored and discontinued at the appropriate time.

Distribution requirements planning (DRP) — System of replenishing distribution by placing time-phased signals on the producing plant. The DRP signals are normally placed as orders in the master production schedule.

DMAIC — Six Sigma problem-solving structure: Define, Measure, Analyze, Improve, Control.

Drop and drag — Functionality within software to move requirements by dragging them to alternative time frames utilizing the computer's mouse.

Effective date — Date that changes (such as engineering changes) to bill of material structure are to take place. These are normally predetermined in advance of the change to use up inventory, except when safety or special service needs are involved in the change.

Enterprise resource planning (ERP) — A methodology of linking and measuring business processes for the objective of high performance and low cost. Emphasis is on capacity and realities of process. It evolved from the MRP II process methodology, but added more emphasis on linkages outside the business (suppliers and customers). Sometimes referred to as supply chain management.

Enterprise resource planning business system model — Process map of the elements and process linkages created by the enterprise resource planning system.

Final assembly schedule — Schedule to drive assembly in an assemble-to-order environment. The master production schedule is normally driving sublevels of the final configuration in this scenario.

Finished goods — Customer-ready inventory.

Finite scheduling — Scheduling very specific requirements to very specific time frames and resources.

Firm fence — Second time fence from the ship date. Although the rules are more tolerant of change between the fixed and firm fences, the objective is still stability if possible. Internal rules of engagement apply.

First-time quality — Measure of the yield through a process in the first pass.

Fishbone analysis — Cause and effect problem-solving tool using a fishbone diagram and process.

5S — Lean Japanese methodology for improving and maintaining housekeeping and workplace organization.

5-why diagram — Problem-solving process for breaking bigger problems into smaller components and determining causes.

Fixed fence — First time fence from ship date. This is usually the time frame where stability is maintained if at all possible. Internal rules of engagement apply.

Forecast — *See* Demand plan.

Forward scheduling — Scheduling using the lead time for sequencing and capacity commitment or allocation.

Independent demand — Demand driven from an outside source such as customer requirements. Not driven from a higher bill of material requirement for the specific case being described. Aftermarket demand for a component is often independent demand when ordered even though the part is resident in another bill of material assembly structure.

Inventory — Material, components, or finished goods held for or used in the process of manufacturing or distribution.

Inventory record — A computer system record showing the quantity of on-hand inventory of any specific item or material.

Inventory strategy — Strategy for where (or if) in the process inventory is to be held in anticipation of an order.

Ishikawa diagram — Cause and effect problem-solving tool using a fishbone diagram and process.

ISO (International Organization for Standardization) — Issues quality and consistency standards for organizations worldwide.

Item master — File where the item master records are stored and maintained. *See* Item master record.

Item master record — The "master" record by part number within the computer system where vital information linked to that item or material is kept.

Just in time — Focus on moving inventory into position at the point of need. Forerunner to lean concepts.

Kaizen — Improvement focus executed in a short period, usually either one day or one week.

Kanban — Method of pulling inventory into the process through the use of a card or other process signal.

Lean — Focus on the elimination of waste in any process. Also associated with movement toward improved flexibility and speed of process.

Level loading — Process of converting demand that may not be level to a more smooth production schedule.

Location — Labeled area where inventory can be or is held. The location description is also sometimes referred to as the location.

Make to order (MTO) — Inventory strategy where either no inventory or some inventory is kept in raw form only until the customer order is received.

Make to stock (MTS) — Inventory strategy where all inventory is kept in finished goods ready for customer demand.

Management systems — Processes in place to provide an accountability infrastructure for sustaining and improving performance. Examples include sales and operations planning and the weekly performance review process.

Manufacturing execution system (MES) — System to aid in the shop floor prioritization of orders and execution.

Manufacturing resource planning (MRP II) — The predecessor to enterprise resource planning.

Master production schedule (MPS) — The top-level detailed schedule that aligns a production facility with customer need.

Master production schedule knowns — Master production schedule requirements driven directly from customer signals.

Master production schedule unknowns — Master production schedule requirements driven from forecasted signals.

Material issue — Decreasing an inventory balance.

Material requirements planning (MRP) — A subprocess of enterprise resource planning that nets available and scheduled inventory against requirements from the master schedule.

Metrics — Measurements used to capture data on process for the purpose of eliminating barriers to high performance.

Netting — Sorting and calculating of changes to the schedule requirements done by material requirements planning.

Obsolete inventory — Inventory that is no longer called out in current configurations and is not planned for service requirements.

Oliver Wight — One of the forefathers of the material requirements planning process.

Operations plan — Capacity plan monitored by top management, usually stated in monthly buckets.

Order configurator — Software tool or manual mechanism to sort customer orders and match to the appropriate internal stockkeeping unit configuration.

Order policy — Policy in materials management governing the rules for order quantities. Examples include lot for lot, fixed order quantity, fixed period quantity, etc.

Pareto chart — Chart for capturing the root causes of process variation. Charted most frequent to least.

Pegging — Requirements attached through the bill of material structure to a record in the system.

Phantom bill of material — *See* Planning bill of material

Planning bill of material — Bill of material used to position material for capacity in anticipation of a customer order.

Pre-S&OP — Final demand review before the sales and operations planning meeting. Questions are answered in preparation for the top management review pending.

Preventative maintenance — Maintenance scheduled to keep machinery in working and reliable condition.

Process map — Tool for mapping the steps of a process for the purpose of evaluating improvement opportunities.

Product family — Grouping of products that share similar traits, such as components, process, and/or inventory strategy.

Product mix — Variations in product demand that require different products to be needed in specific time periods.

Productive maintenance — Similar to preventative maintenance, but normally the machine operators are involved in portions of the servicing, as are maintenance personnel.

Quad chart — Performance reporting chart that includes performance, trends, root cause, and actions/name/dates for action. Used in the weekly performance review.

Remanufacturing — Making used product into saleable product for the rebuilt sales market business.

Reverse auctions — Web-based auctions where suppliers bid for the privilege of business by bidding down their supplier price or upping the bid for services to be supplied to a specific customer.

Root cause — Underlying reason for process variation. Normally looked at as an opportunity for improvement.

Routing record — Documented description of the process steps used in manufacturing or converting raw material and/or components into finished and semi-finished goods.

Sales and operations planning (S&OP) — The top management planning process where risk management is exercised, especially in the cases of new product introduction, transfers of product, and capacity concerns as they relate to demand expectations.

Seasonality — Differences in product demand based on somewhat predictable seasons of the market.

Setup — Time, resource, and effort spent from last good piece to new good piece in a process changeover.

Single minute exchange of die (SMED) — Process methodology for reducing changeover time. Also applies to processes such as preventative maintenance.

SIOP — An acronym with the same meaning as S&OP (*see* Sales and operations planning). Stands for Sales, Inventory, and Operations Planning.

Six Sigma — Improvement focus based on the elimination or reduction of defects.

Software — Computer program that allows consistent process to be executed in a specific area of focus, such as enterprise resource planning business system software.

Supply chain management — Controlling of component feeds into the manufacturing process. Normally includes capacity planning, supplier analysis, and measurements/feedback.

Takt time — Rate at which the customer demand requires product. The master scheduler needs to match this against production capacity. Stocking decisions and production rates are determined from this information.

Theoretical capacity — The capacity of a process or machine when very little or no process variation is present. Not used in high-performance master scheduling except as reference only.

Time fence — Positions in the accumulative lead time of a product where rules change or decisions are made. Policies and procedures refer to these time points.

Transaction — A record of action (labor completed, movement of material, etc.) in the business system.

Two-level master schedule — Master production schedule where sublevel components or assemblies are planned in anticipation of the purchase order. The final assembly schedule is the "top" level of this schedule.

Unit of measure — Descriptor for volume labeling such as feet, inches, each, pounds, grams, etc. Linked to the bill of material records in a business system.

Vendor-managed inventory — Inventory or flexible capacity in the supply chain positioned in anticipation of demand from the next level in the bill of material.

Visible factory board — Factory measurement board on which department performance is posted.

Waste — Nonvalue-added time, money, capital, or other resource that is not utilized for gain.

Weekly performance review — Weekly accountability infrastructure for reviewing the preceding week's performance in schedule and data accuracy.

Weekly project review — Scheduled weekly management review of open projects.

Work in process — Inventory issued to the manufacturing process.

Yield — Rate at which product is successfully processed or manufactured to specification.

CHECKLIST FOR HIGH-PERFORMANCE MASTER SCHEDULING

1. Top management engages in a top management sales and operations planning (S&OP) process.
2. The master production schedule (MPS) is directly linked to the S&OP process in step 1.
3. The master scheduler is actively involved in the S&OP process.
4. The MPS is updated frequently to align with current demand requirements. It is realigned to current resources.
5. The MPS has no past due requirements at the start of a period.
6. The MPS horizon is at least 12 months rolling.
7. The MPS is linked directly to the material planning process.
8. There is a direct feedback loop from procurement and shop floor control to update issues affecting the MPS performance.
9. The MPS is measured daily, weekly, and monthly.
10. A separate MPS is done for each product family
11. Product families reflect product commonalities, including constraints, common components, and inventory strategy.
12. The MPS is scheduling the proper level for each inventory strategy (finished goods for make to stock, subassemblies for assemble to order, raw materials and components for make to order, etc.).
13. Time fence disciplines are administered and maintained.

14. Rules of engagement between order promising and scheduling are well understood, documented, and disciplined.
15. Inventory strategies are reviewed at least quarterly.
16. Process constraints are acknowledged and managed.
17. Planning bills of material are utilized for planning "unknowns."
18. The master scheduler is accountable for MPS accuracy.
19. The master scheduler participates in a regular (weekly) MPS metric review process for root cause and continuous improvement.
20. MPS and materials planning performance are both at or above 95% accuracy.

INDEX

A

ABC stratification, 75
Accountability infrastructure, 182
Accounting, 197
Accumulative lead time, 5, 13, 18, 25
Accuracy. *See* Demand planning
Advanced planning and scheduling, 157–158
Alignment, 135
AlliedSignal, 174, 191, 213
American Production and Inventory
    Control Society (APICS), 170–171,
    226
  certification, 164, 219
Assemble to order (ATO), 17–18, 23–24
  fixed fence rules in, 88–89
ATP. *See* Available to promise
Audit, 179–182, 183
Autopilot, 14, 208, 216, 224
Available to promise (ATP), 59–63, 87,
    208, 217, 224–225

B

Baker, Eugene, 171
Barometric measurements, 127
Best Buy, 17
Bill of material (BOM), 7–8
  accuracy, 35–36, 73
  planning, 8, 68–71

product families and, 66–68
time frames for planning, 70
unique components within, 69
Blanket orders, 55–57
Blow-molded bottles, 20
BOM. *See* Bill of material
Brainstorming, 194
Buffer
  inventory plan, 4–6
  "required plus" level, 69
  as top management decision, 68–70
  "unknowns" and, 33
Buker, David, 171, 172
Business planning, 44–46

C

"Can-do-check-act" wheel, 192
Capability, 91
Capability measurement, 188
Capacity, 97, 99, 205, 214, 222
Capacity planning, 8–11, 143
Capital equipment, 49
Capital goods market, 85
Cash flow, 197
Caterpillar, 77
Cause and effect diagram, 196, 197
Cause and effect with the addition of
    cards (CEDAC), 197
Centralized master scheduling, 79–81

239